Propellor Aerodynamics

The History, Aerodynamics & Operation of Aircraft Propellers

Frank E. Hitchens

Andrews UK Limited

Copyright © 2015 Frank E. Hitchens

The right of Frank E. Hitchens to be identified as author of this book has been asserted in accordance with section 77 and 78 of the Copyrights Designs and Patents Act 1988.

All rights reserved. No part of this publication may be reproduced, stored in a retrieval system, or transmitted in any form or by any means, without the prior permission in writing of the publisher, nor be otherwise circulated in any form of binding or cover other than that in which it is published and without a similar condition including this condition being imposed on the subsequent publisher.

First published worldwide by
Andrews UK Limited
The Hat Factory
Bute Street
Luton, LU1 2EY
www.andrewsuk.com

Front cover photo: Lockheed P3-C Orion propeller.

Contents

Introduction vii

1 – The History of Aircraft Propellers

In the Beginning…. 1	Lockheed C-130 Hercules . 19
Airscrews 1	De-ice for Props 20
The First Airborne Props . . 2	Propeller Manufacturers. . 21
The Wright Brothers 3	The First Turboprops. . . . 24
Aerodynamic Theories 5	Supersonic Propellers . . . 26
Variable-pitch Propellers . . 6	Record Breakers 28
Constant-speed Propellers . 8	Propulsors. 31
Spitfires and Hurricanes . . 11	Propfans 32
Five-blades or Six?. 11	Conclusion. 35
Counter & Contra-rotating Props. . 14	

2 – Propeller Pitch

The Purpose of the Propeller 37	Slip 51
The Working Fluid. 37	Brief Review 52
Propeller Terminology . . . 39	Advance/Diameter Ratio. . 54
Propeller Pitch 42	Fixed-pitch Propellers . . . 56
Helix Angle 43	Variable Pitch & Constant Speed Propellers 59
Blade Angle & Twist. 45	
Experimental Pitch 45	
Geometric Pitch 47	

3 – Thrust & Efficiency

Efficiency	71	Number of Blades	86
Propulsive Efficiency	74	Aspect Ratio	91
Power to the Prop	77	Prop Blade Loading	93
Power Absorption	82	Prop Disc Loading	94
Activity Factor	84	Propwash Thrust	96
Solidity	85	Prop Blade Drag	103
Prop Diameter	85	Propeller Icing	108

4 – Effect on the Aircraft's Stability

Prop Torque Force	113	Centreline Thrust	124
Prop Location	114	Minimum Control Speed (V_{MC})	125
Helical Propwash	118		
'P' Factor	119	Counter-rotating Propellers	126
Gyroscopic Effect	121	Contra-rotating Propellers	129

5 – Prop Tip Speed & Noise

Tip Speed & Noise	131	Tips, Blade Shape & Materials	137
The Cause of Noise	131		
High-speed Aerodynamics	132	Synchronizing	142
Tip Speed	135	How Noisy are They?	143

6 – Propeller Forces & Stress

Prop Forces in Cruise Flight	147	Prop Stress	153
Windmilling Prop Forces	148	The Balanced Prop	161
Reverse Thrust Forces	151		

7 – Turboprops, Propulsors & Propfans

Turboprops	163
Propulsors	166
Propfans	169

8 – Constant-speed Units

The Constant-speed Unit. 173
The Oil/Counterweight Type 176
The Beech & McCauley Types 176
The Hamilton Standard Type 177
Air/Oil Type. 178
The Hydromatic CSU . . . 178
The Curtiss Electric Propeller 179

9 – Propeller Operation

Magneto Drop &
   Leaning the Mixture. . 183
Governor Check 183
Running Square 185
Reducing Power 186
Lock-on. 187
Overspeed Condition . . . 188
Windmilling. 189
The Feathering Prop . . . 189
Ground Feathering 193
Negative Torque System . 195
Autofeathering 195
Simulated Zero Thrust . . 196
Reverse Thrust. 197
Safety Around the Prop . 198
Conclusion. 204

Glossary **207**

Bibliography **217**

Index **219**

Introduction

Imagine you are the pilot of a light aircraft on a cross-country VFR flight. As you tick-off each landmark along your route, you watch as it disappears below the leading edge of the wing to reappear a few seconds later behind the trailing edge. Your thoughts turn to your ground studies and knowledge of basic classical aerodynamics. You understand how the airflow separates at the wing's leading edge and flows over and under the wing to rejoin at the trailing edge. You also understand the forces of lift and drag produced by the aerodynamic reaction of the aircraft's wing. Now, as you look ahead for the next approaching landmark, your view is through the almost invisible blur of the propeller disc. Because the prop is an almost continuous blur, we tend to ignore its presence and take it for granted.

Therefore, what about the propeller's forces of thrust and torque, or the prop's stress, tip speed, power absorption and it's efficiency of operation, etc? Do you ever think about them? Where did propellers originate? They have been around a great deal longer than you may realise! How do you operate a constant speed or feathering propeller? How do props do their job of producing thrust? That is what this book is all about – propeller aerodynamics, covering the history of the propeller's development, its operation and of course, the aerodynamics associated to the propeller.

Most text books on general aerodynamics give only a brief mention to propeller aerodynamics. It has been this author's intent to present this material in an easy to understand manner, suitable for study by the low time private pilot. Nevertheless, that's not to say the more knowledgeable reader won't benefit from this book. It has been assumed the reader has an understanding of basic aerodynamics to at least the private pilot's level. The text on propeller aerodynamics will therefore, compliment his/her knowledge on this fascinating subject.

We start with a look at the propeller's history of development and followed by the different aerodynamic theories put forward by William Rankine, Robert and William Froude and Stefan Drzeweicke, which in this book, concentrates mainly on the blade element theory and briefly, on the axial momentum theory. The text continues with different aspects of propeller pitch and the factors that affect the propeller's efficiency. This is followed by the forces acting on the propeller during different operating conditions and is followed by a brief look at turboprops, Propulsor and Propfans. The book concludes with a chapter on propeller operation. A few simple formulas have been included along with several diagrams to help clarify the text. Note, all diagrams have been drawn free hand by this author and then computer scanned and do not represent any on particular propeller or airplane.

In writing this book, a choice had to be made on the use of either Imperial or Metric units. With my home country of New Zealand and many other overseas countries turning to the Metric system more and more, this was at first deemed to be the most appropriate system to use. However, in the aviation industry, Imperial units are still commonly used. For example propeller size and manifold pressure are still measured in inches, and piston-engine power mostly in brake horsepower, true air speed in Knots, prop thrust in pounds, and prop disc area in square feet. One notable exception being temperature measured in the Metric system of degrees Celsius. Young pilots may well be familiar with the Metric system while older generation pilots (myself included) will be more familiar with Imperial units, which I have chosen to use in this work with Metric equivalents in brackets.

A total of sixty-one photographs are included from this author's collection with the exception of the MD-80 Propfan test plane photograph, which was freely donated by Hamilton Standard of Connecticut, USA, to whom I am truly thankful. My thanks also go out to those pilots and ground engineers who were most helpful in supplying useful information and answering my numerous questions on aircraft propellers and arranging access to aircraft parked in the restricted operational areas of the airport I visited.

Frank Hitchens,

Wellington, New Zealand.

1 – The History of Aircraft Propellers

In the Beginning...

Before the advent of jet propulsion, a piston-engine driving a propeller to provide the necessary thrust or forward motion powered all aircraft. To this end, the propeller has always been an accepted part of an airplane. However, how many people realize that propellers were around long before the first airplane flew? So, where did propellers originate? Some authorities claim the propeller originated in China several centuries ago as a descendant from the windmill. In Europe, the windmill can be traced as far back as the 13th century AD, but windmills built in China before this time were of a different type, their axles being vertical. It is believed the windmills of China have no apparent relationship with European windmills, however, from the principle of the windmill the idea of the propeller was born.

Airscrews

The word 'airscrew' was introduced to aviation to distinguish between the aeronautical and marine type propellers (which were usually referred to as 'screws'). The word airscrew was more commonly used in Europe than in the USA. Between the 1920s and 1950s, the name airscrew usually referred to a 'tractor' propeller (a propeller in front of the engine as opposed to pusher propeller behind the engine). It is now a virtually redundant term and has been replaced by the word propeller or prop for short, although the word airscrew is quite often used by writers of early aeronautical history when writing about propellers of that era. The term propeller was first used to describe any

1 – The History of Aircraft Propellers

The term 'airscrew' is synonymous with early aircraft such as this example of a Sopwith F.1. Camel Scout.

mechanical device used to propel a vehicle and it came into aviation terminology circa 1850 to have the same meaning as the word airscrew.

The First Airborne Props

At the end of the 15th century, circa 1490, the artist and inventor Leonardo Da Vinci (1452–1519) designed a crude form of airscrew (or prop) for his helicopter design. The design never left the drawing board but the word helicopter has been with us ever since. The French mathematician, J.P. Paucton introduced the idea of using two propellers on airships, one to propel the craft forwards and the other to lift it upwards; it was not successful! It was to be nearly three hundred years until 16 October 1784, when Jeanne-Pierre Blanchard (1753–1809) used the first airborne propeller on his hot air balloon. The propeller consisted of three metal plates attached to the end of poles, which were rotated by hand and again, were not successful. On 24 September 1852, Henri Gifford (1825–1882) a French engineer, used a three horsepower (2.24 kW) steam engine on his dirigible (steerable) air ship driving a three-blade 11 feet (3.35 m) propeller to achieve the first powered flight covering a distance of 18 miles (30 km) from Paris to Trappes. On 9 August 1884, Captains Charles Renard and Arthur Krebbs of the French Corps of Engineers completed a circular coarse of five miles (8 km) in their airship *La France*. The 23 feet (7.01 m) four-blade wooden propeller was powered by a nine horsepower (6.7 kW) Gramme electric motor. The prop turned at a very slow 50 RPM. Around this time, other pioneers had varying amounts of success, propelling their airships using propellers turned by hand, or powered by electric or petrol driven engines.

The Wright Brothers

It is now a well-known fact the Wright brothers are credited with being the first to achieve sustained, powered flight in an airplane using their Wright Flyer I, on 17 December 1903. The flight took place at Kittyhawk, North Carolina, USA, with Orville Wright (1871–1948) at the controls. Several other pilots from different parts of the world also claim to be the first to fly, before the Wright brother's famous first flight. However, the Wright's first flight has long been recognized as the first successful 'controlled and sustained flight' in heavier than air aircraft, which the other pilots failed to achieve.

The Wright's success was achieved through having a suitable engine and propeller combination. Both were of course, designed and made by them with the assistance of their mechanic Charles Edward Taylor (1868–1956. Their propeller design calculations were remarkably accurate, with the prop twist being correct for the speed ratio of their props. When they designed their props, they had to rely on their own calculations because not many (if any at all) calculations were available on aircraft propellers back in those days of early flight. As we shall see shortly, various theories were available for ship's propellers which have since been applied to aircraft propellers. Each prop at 350 RPM produced sixty-seven pounds of thrust. The propellers were hand carved from three laminations of spruce wood and painted with aluminium to prevent their work from being copied by any competitors. It is not known which of the two Wright brothers carved the props; the design was never patented. Because the props were contra-rotating, it was not a simple matter of making the second prop a copy of the first one; it had to be a mirror image due to rotating in the opposite direction. Their mechanic Charles Taylor built the 12 horsepower (9 kW) four-cylinder engine, which turned by train drive the two 8.5 feet (2.59 m) propellers. One chain was twisted into a figure of eight loop to drive the second prop in the opposite direction to the first. Tests conducted at the NASA Langley Full Scale Wind Tunnel, revealed an efficiency of 81.5% on the Wrights' 1911 propellers, a remarkable achievement for that era. Their efforts were a great contribution to the development of aircraft and propellers that have since followed.

Normal stress forces acting on the propeller are to be expected and are allowed for in the modern propeller's design but undue stress can lead to disaster. In fact, the first person to be killed in an aircraft accident was due to propeller failure. Lt Thomas Selfridge was killed while flying with Orville Wright in 17 September 1908. Orville was severely injured but survived.

Aerodynamic Theories

Several theories on propeller aerodynamics have been put forward by different authors. The theories include the vortex, general momentum, axial momentum and the blade element theory. The blade element is the most widely accepted theory, along with parts of the axial momentum theory, leaving the vortex and general momentum theories to fall by the wayside.

In 1865, the Scottish engineer and scientist William George Rankine (1820–1872) founded the axial momentum theory while working on the theory of ship's propellers. At a later date further work and development on the axial momentum theory, was covered by Robert Edmund Froude (1846–1924) also an engineer. The blade element theory was first introduced by William Froude (1810–1879) the father of R. E. Froude, an engineer and naval architect in 1878, when he also was working on ship's propeller theories. Note the theory of ship and aircraft propellers is virtually the same, because air at subsonic speed behaves very similar to flowing water.

Stefan Drzeweicke (1844–1938) further developed the blade element theory from 1892 onwards and he has been credited with the majority of the research work. The axial momentum theory, also known as the Rankine-Froude theory after its two authors, deals with the energy change given to the air mass after it passes through the propeller disc. It also includes the effect of the rotational propwash, the friction drag of the propeller blades and the loss of energy in the propwash caused by the interference of the engine nacelle or the fuselage, amongst other factors. The blade element theory deals with the forces acting on the propeller as it moves through the air at a uniform velocity. It also includes the blade's shape and number of blades and assumes the propeller blade to be made from an infinite number of blade elements, hence the name blade element theory. Theodore Theodorsen (1897–1978) of NACA also performed aircraft propeller research, circa 1930s. The diagrams in this book represent the blade element theory.

After the Wright brothers' initial success, further testing and advances in prop design by other engineers led to the first generation of propellers. Lucian Chauviere (1876–1966), a French aeronautical engineer, is noted for introducing his *Integrale* aircraft propellers of advanced design, (he introduced the laminated propeller mentioned above) the forerunner of all propellers to follow. Louise Bleriot (1872–1936) had a Chauviere propeller mounted on his Bleriot XI monoplane for his cross-channel flight in 1909. During the First World War, Chauviere's company produced about 25% of the propellers manufactured for allied aircraft.

Variable-pitch Propellers

Fixed-pitch props are quite suitable for low-speed light aircraft with low horsepower. However, as engines with greater horsepower were developed producing greater cruise speeds, the fixed-pitch props suffered in performance. Thus problem was resolved when the next important step in propeller history occurred in 1916 when a variable-pitch (VP) propeller with forward and reverse thrust was first used on an airship to help with manoeuvring. The idea of a VP prop was first mooted by the Frenchman Croce-Spinelli as early as 1871 without any progress in that area. The first variable-pitch propeller mounted on an aircraft was flight-tested on 23 October 1922 by Sandy Fairchild; this was a VP propeller made by the American Propeller Company, with two positions for forward and reverse flight. Nevertheless, the major part of the prop's development was performed in 1926 in the United Kingdom at the Royal Aircraft Factory in Farnborough (now known as the Royal Aircraft Establishment or simply as Farnborough). A Royal Aircraft Factory RE-8 airplane was used to flight-test the four-blade, wooden VP propeller. The flight-testing of VP props was also progressing well in Germany, with the German R-36 bomber being the first multi-engine aircraft to fly with twin VP props. Variable-pitch props were used on many aircraft types between 1928 and 1940. The propeller manufacturer Standard Steel in the USA made the ground adjustable VP propeller for Charles Lindbergh's Ryan NY-P 'Sprit of St. Louis' in which he achieved the first solo transatlantic flight on 20–21 May 1927.

The US Army aircraft Engineering Division at Wright Field, Dayton, Ohio, further developed VP props under the direction of Frank Caldwell. He later joined Hamilton

A replica of Lindbergh's Ryan NYP with a ground adjustable prop, located in the San Diego Air & Space Museum, Ca.

Standard Division of the United Aircraft Company as their Chief Engineer, where he perfected controllable pitch propellers in 1933 for which he received the Collier Trophy.

Reverse thrust props were not operational until 1943 when they became popular for manoeuvring flying boats on the water. Two years later reverse thrust found its way onto conventional, wheeled aircraft to assist landing roll braking. One of the first aircraft to be so equipped was the Boeing B17 Flying Fortress bomber. Just about all modern turboprop aircraft are now equipped with reverse pitch propellers. It is rare to find a reverse pitch prop on a single-engine aircraft but not altogether unknown. Although VP props have remained in limited production over the year's right up to the present day, its place has been taken over by the introduction of the constant-speed propeller for high-performance aircraft.

Constant-speed Propellers

The constant-speed propeller was patented in 1924, in the UK, by Gloster/Hele/Shaw/Beacham. Flight-testing was carried out on a Gloster Grebe in 1926–7. The propeller built of compressed wood was later produced by Rotol, now Dowty Rotol in the UK. Although the constant-speed prop was developed in the UK, the American built Grumman FF-1 is believed to be the first production aircraft to use a constant-speed prop. This was a single-engine biplane fighter, which entered service with the USAAF in late 1931.

It was to be another ten years before the Royal Air Force operated aircraft with constant-speed propellers. In 1929, another propeller manufacturer known then as Hamilton Aero, in the USA, merged with Standard Steel to form the now well-known Hamilton Standard

A constant-speed propeller mounted on a Boeing B-17G Flying Fortress. The chrome dome on the hub houses the constant-speed unit piston. Pima Air & Space Museum, Tucson, Ar.

company. On 10 June 1999, Hamilton Standard's parent company United Technologies, acquired the Sundstrand Corporation merging it with Hamilton Standard to form Hamilton Sundstrand. During the 1930s, Hamilton Standard designed a constant-speed, three-blade prop made from a light alloy, the pitch-change mechanism being hydraulically operated. This was first used on the Boeing 247D, to become the first modern airliner to use a constant-speed prop, in 1934. The three-blade Hamilton Standard prop was powered by a geared Pratt & Whitney Wasp radial engine of 550 BHP (410 kW). This prop replaced a direct drive Wasp turning either a two-blade or a three-blade variable-pitch propeller found on earlier models of the Boeing 247. The improved performance given by the constant-speed propeller enabled the Boeing 247D to become the first airliner capable of climbing on one engine.

Hamilton Standard invented the 'paddle blade prop' circa 1940, to produce greater static thrust for better take-off performance. Paddle blade props have broad constant chords, usually with blunt tips on modern propellers. Hamilton Standard also introduced the 'Hydromatic' feathering propeller, which went into service in 1936, and during World War 2, Hamilton Standard went on to make propellers for three quarters of the allied aircraft.

The first company in the USA to manufacture propellers was the Regina Gibson Company in 1909, which was placed in the hands of the Canadian engineer Wallace R. Turnbull. An alternative to the hydraulic constant-speed unit is the electrically operated system, first designed by Turnbull in 1925 and followed by flight-testing in 1927. Curtiss-Wright in the USA licensed and improved on Turnbull's design and made aircraft propellers with electrically operated pitch change mechanisms. The Curtiss Electric propeller was very popular during WW II, being used on many types of American built single-engine fighters and light twin-engine bombers. Because of necessity for high-performance aircraft, the constant-speed propeller is assured a long and very productive future.

Spitfires and Hurricanes

The Supermarine Spitfire of Second World War fame first flew with a two-blade, fixed-pitch, wooden prop during 1938–9. By the early stages of the war in 1940, the two-blade propeller had been exchanged for a metal three-blade, constant-speed prop. In 1942, the Spitfire Mark IX was flown with a four-blade propeller to absorb the increased brake horsepower of the more powerful engines that followed. Late model Spitfires with the Griffon 65 engine or similar, were powered by a five-blade prop and later still, with a six-blade contra-rotating propeller.

The Spitfire's stable mate, the Hawker Hurricane was similarly equipped with the early models flying with a Watts, two-blade, fixed-pitch wooden prop. In 1939–40, the two-blade prop was swapped for a three-blade metal, two-position VP propeller. Canadian built Hurricanes with the Merlin III engine powered a DH-Hamilton Standard Hydromatic, three-blade propeller.

Five-blades or Six?

As mentioned above, it became necessary over the years to increase the number of blades mounted on a prop to absorb the greater horsepower of newer types of engines. Four-blade props became more common and still are, and later five-blade props were produced. Due to interference caused as the air flow cascades over the following blades,

The Supermarine Spitfire Fr XIVe has a five-blade propeller. Manchester Air Museum, England, is the home to this aircraft.

The Lockheed Hercules C130J with six-blade composite propellers. The Original C130A model had three-blade props.

it was found that five-blades were the maximum that could be used before propeller efficiency deteriorated. However, where the metal prop was limited to five-blades, props with six-blades are now powering modern turboprop transports and cargo planes. Composite materials enabled six-blades to be mounted on each propeller hub, generating greater thrust more efficiently than a metal prop could ever achieve. For example, the BAe ATP and later models of the Aerospatiale ATR and also the latest Lockheed Hercules C130J all have six-blade composite props, to name just a few.

To this author's knowledge, the only six-blade metal prop to fly (not counting contra-rotating props) was a pusher prop mounted on the Japanese Kyushu J7W1 Shinden that was being developed at the end of the Second World War, but never saw combat service.

1 – THE HISTORY OF AIRCRAFT PROPELLERS

The counter-rotating propellers on this Lockheed P-38 Lightning are clearly visible. This aircraft is located in the Udvar-Hazy Center, near Washington D.C.

Counter & Contra-rotating Props

On twin-engine aircraft, the propellers of each engine may rotate in opposite directions (counter-rotate) to enhance engine out handling. The idea of using counter or contra-rotating propellers is not a new one. The Wright brothers' Flyer I built in 1903 had counter-rotating propellers, chain-driven off one engine. The chains were not bicycle chains as some people believed, but were made especially for the Flyer I.

The 11-feet (3.35 m) Curtiss electric props on the Lockheed XP-38 Lightning (1939) experimental fighter rotated top blade in. On the YP-38 Lightning prototype, and all subsequent production models, the prop's direction of rotation was reversed to rotate top blade out. The result was exceptional longitudinal stability for the P-38.

The Piper Twin Comanche and its bigger brother the Navajo Chieftain, both have counter-rotating props rotating top blade in towards the fuselage, the conventional way for modern aircraft. An exception was made with the Piper Aerostar 700P; this aircraft has its counter-rotating props turning top blade away from the fuselage, the same direction as on the P-38 and also for stability purposes. Because the blades rotate top blade outwards, the centre of thrust is placed further outboard but the critical engine is still eliminated.

Another aircraft of historical significance with counter-rotating propeller is the Shorts S.39 Triple-Twin. The name sounds contradictory until one realizes the layout of the aircraft; the two seat aircraft looks similar in some respects to the Wright Flyer I, with a wing span of 34 feet (10.4 m) and a length of 45 feet (1.7 m). It had two Gnome rotary engines of 50 BHP (37 kW) each, driving three propellers, hence the name Triple-Twin. The engines were mounted in tandem in the fuselage; the front engine powered two tractor, counter-rotating propellers by chain drive, similar to the Flyer I, while the rear mounted engine powered a single pusher prop. The Triple-Twin first flew on 18 September 1911, to become the world's first aircraft driven by three propellers, albeit with only two engines, the forerunner of the push/pull arrangement found on later aircraft. On 13 May 1913, the flight of the world's first four-engine aircraft occurred; this was the Bolshoi Baltiski (Grand Baltic 1) developed by Igor Sikorski. First designed as a twin-engine biplane that was underpowered, two more engine and propellers were added in tandem with the original Argus engines, each of 100 BHP.

A contra-rotating propeller (or contra-prop for short) is one that consists of two co-axial mounted propellers and driven by the same engine, but rotating in opposite

1 – The History of Aircraft Propellers

directions. With this arrangement, the total amount of power absorbed by the propeller can be greatly increased. Contra-props also date back to the early days of flying. At least as far back as 1909 when the Piggott brothers built their Piggott biplane with a contra-prop. [There could have been others before this one, but not to this author's knowledge]. The two props on the Piggott were well spaced being about 3 feet (1 m) apart. Conventional contra-props are placed adjacent to each other as shown in the photograph of the Avro Shackleton.

A Deperdussin with a contra-prop followed the Piggott's biplane in 1912. Moreover, 1944 saw the appearance of the experimental Boeing XF8-B1 carrier-based fighter/bomber powered by a contra-prop. It never passed the experimental stage and so, did not make it into military service.

The Avro Shackleton bomber has four engines each powering a set of contra-props. This aircraft is in the Manchester Air Museum, England.

COUNTER & CONTRA-ROTATING PROPS

A Supermarine Seafire F.17 with a four-blade propeller, similar to the Seafire mentioned in the text. This aircraft is located in the Fleet Air Arm's Naval Air Station Air Museum, Yoevilton, England.

In 1944–45, a Supermarine Spitfire XIV was flight tested for 230 hours with a six-blade contra-rotating propeller installed. The first flight was performed in August 1944 and was followed by the carrier based naval version of the Spitfire – the Seafire 47. A Rolls Royce Griffon engine of 2350 BHP driving a six-blade contra-prop powered the experimental Seafire. This was a substantial increase in power from the early 'Mark' of Spitfire's engines of 1000 BHP driving a two-blade propeller.

Four years later, The UK's Fleet Air Arm's Fairey Gannett prototype became the world's first contra-prop equipped aircraft to be powered by a turbine engine – the Armstrong Siddeley Double Mamba 2950 ESHP turboprop powered the eight-blade contra-rotating propeller. The Gannett's first flight took place on 19 September 1949. It went on

1 – THE HISTORY OF AIRCRAFT PROPELLERS

An example of the AEW-3 version of the Fairey Gannet. The Pima Air & Space Museum in Tucson, Arizona, is home to this aircraft.

to make further aviation history on 19 June 1950 when it became the first turboprop aircraft to make a carrier landing; the ship was the HMS *Illustrious*.

The prototype Bristol Brabazon piston-engine airliner was another contra-rotating propeller equipped aircraft worthy of note. Its eight Bristol Centaurus engines each producing 3500 BHP, were coupled in pairs; each powered an eight-blade Rotol contra-prop. After 400 hours of flight-testing over a period of four years, the prototype was scrapped. Its demise was initiated by the high cost of development and the fact the DeHavilland Comet 1 jet airliner made its maiden flight about six weeks before the Brabazon's first flight, which was accomplished on 4 September 1949. The end of the large piston powered airliners had finally arrived; jets were here to stay!

Moving on to larger aircraft... The Russian built Antonov AN-22 Antheus (1965) was for a short time the world's largest aircraft until the arrival of the Boeing 747 and the Lockheed C-5A Galaxy military transport. The An-22 is powered by the world's largest turboprop engines; these are the Kuznetsov NK-12MA fixed-shaft engines developing an incredible 15,000 Shaft Horsepower (11,185 kW) each. The four engines each drive massive 20 feet 4 inch (6.2 m) eight-blade contra-props at a very low 750 RPM, and producing a very distinctive noise in the process. As a further point of interest, the AN-22's props are very large but they are not the largest propellers to be used. Although it was not a contra-prop, the record for the world's largest propeller goes to the 22 feet 8 inch (6.9 m) Garuda prop. The Garuda was mounted on the German designed, Polish built Linke-Hoffmann R-11 prototype aircraft, with a first flight on 19 January 1919. It looked like a single-engine aircraft on steroids, but had a wing span of 138 feet 4 inches (42.25 m) and a length of 66 feet 8 inches (20.2 m). Compare this to the wing span of an early Boeing 737 at 93 feet (28.35 m). The single Garuda prop was powered by four Mercedes D.IVa of 260 BHP engines each, giving a maximum speed of 81 MPH. Due to the propeller's large diameter, it turned at a very low 545 RPM. Only the one prototype was ever completed.

Lockheed C-130 Hercules

It is interesting to note, the Lockheed YC-130A Hercules prototype and also the first ten production models were equipped with Curtiss-Wright, three-blade, electric propellers. Powered by the Allison T56 turboprop engines, the props rotated at a constant 1108 RPM, the thrust was adjusted by the prop's blade angle. However, a problem with the electrically operated governor caused the CSU to overcompensate or hunt in either direction. This resulted

1 – THE HISTORY OF AIRCRAFT PROPELLERS

The Lockheed C-130D Hercules was powered by three-blade propellers. This aircraft is a resident of the Pima Air & Space Museum, Tucson.

in uneven thrust being produced and caused the aircraft's nose to yaw from side to side. On occasions the hunting was so bad the propeller's reduction gearbox would overheat and cause sever damage requiring an in-flight engine shutdown and the propeller to be feathered. The problem with the Curtiss-Wright props was eventually corrected, but not before a change was made to hydraulic CSU's made by Aero Products, Allison's subsidiary company. Subsequent Hercules models from the C-130B onwards were fitted with Hamilton Standard three-blade, and later four-blade propellers. In addition, to follow the trend of modern turboprop aircraft, the latest C-130J models have a new type of turboprop engine, an Allison AE 3100 D3 flat rated to 4591 SHP, driving six-blade composite props.

De-ice for Props

From the early days of flying, ice build-up on the wings and propeller blades has always been a serious problem. Not only is the extra weight a problem, but of greater consequence is the change in the leading edge profile, which alters the airflow over the wings and prop blades, having a detrimental effect on the aircraft's performance characteristics beyond acceptable limits. To help alleviate wing icing, in 1934, the American company B. F. Goodrich pioneered the system of pulsating rubber boots on the wing's leading edge to break off the ice. Ice protection for the prop blades followed later, in the form of de-ice boots, electrical heater mats and a chemical slinger system.

Propeller Manufacturers

Early propellers were made from one piece of solid wood and later props were made from several laminations of a hardwood, usually spruce or mahogany and then hand carved. Shortly after World War I, the first successful metal props appeared; this was the 'Reed' type developed by Dr. S. A. Reed. The props were cut, shaped and twisted from one piece of Duralumin, an aluminium alloy. From 1926 onwards, metal props became more common on high-powered engines, but wood props are still used today, usually on low-powered home-built aircraft.

Ernest G. McCauley (c1885–c1930) founded the McCauley Propeller Company in 1938. Right from the inception of the company, all their propellers were made from metal, starting with a ground adjustable prop they invented and made from solid steel, in 1938. In 1946, they made the first metal props for civilian light aircraft, called a 'Met-L-Prop'. Metal props are now made from either a single piece of forged Duralumin, hollow steel or a light alloy. Prior to 1930, most wooden propellers consisted of either two or four-blades. A three-blade prop was unusual due to the difficulty of attaching the blades to the prop hub, while still maintaining sufficient strength. With metal

1 – THE HISTORY OF AIRCRAFT PROPELLERS

The Junkers JU-88 bomber with three-blade, wooden propellers. The National Museum of the USAF, Dayton, Ohio, is the home for this aircraft.

props this is of course, less of a problem. One notable exception was the Junkers JU-88 twin-engine bomber. Apart from being the most-produced German bomber of the Second World War, it also had three-blade, wooden, constant-speed props. The Cessna Aircraft Company purchased the McCauley Aviation Corporation in 1960 and operated it as a separate division. The same year saw the introduction of McCauley's two-blade, constant-speed propeller with full feathering and alcohol de-ice system. Electric de-ice followed in 1967.

As the years progressed, further advancements saw the introduction of three-blade, full feathering props and reverse pitch props for turboprop aircraft in 1977. Next up was the four-blade prop in 1983 followed by the hugely successful McCauley 'Blackmack' series of propellers. In 1992, the five-blade scimitar shaped prop was made for turboprop aircraft of up to 1650 shaft horsepower. The company received a name change in 1996 to McCauley Propeller Systems and it is now the largest propeller manufacturer with over 250,000 props made over the year, operating on aircraft worldwide.

In 1911, Robert Hartzell expanded his family's Walnut Furniture manufacturing company to make aircraft propellers with the name changed to Hartzell Walnut Propeller Company of Ohio, USA. The name was later shortened to the Hartzell Propeller Company. They concentrated on the high-performance light aircraft and turboprop market; they no longer make fixed-pitch propellers. They were the first company to manufacture feathering propellers and pioneered the use of composites materials in 1945 in propeller manufacturing. Their first composite propeller (and the world's first) was patented in 1949 and flight-tested on a Republic Seabee amphibian aircraft. In 1978, Hartzell installed the first production run prop made totally from composite materials on the Spanish CASA 212c Aviacar. Composite prop blades are 25–50% lighter than metal blades with the added advantage of higher strength, greater reliability and performance, fatigue resistance and better vibration damping, etc. The blades are made from polyester or epoxy-resin with a fibre to provide directional strength. The fibre may be glass, carbon or a synthetic Aramid fibre such as Kevlar, which is also found in bulletproof vests, amongst other uses. It is also lighter in weight and more expensive than ordinary fibreglass. In addition to using new materials, propellers are still being refined with new designs using 'Q-tips', sweptback tips and scimitar shaped blades.

The Sensenich Propeller Company was founded in the USA by the Sensenich brothers in 1932. By 1942, it was the largest manufacturer of wooden propellers in the USA and commenced fixed-pitch metal propeller manufacturing in 1947 and later composite propellers. McCauley, Hartzell and Sensenich are the three leading propeller manufacturers in the USA for light to medium size aircraft. So, is it by chance or coincidence that two of the world's major propeller manufacturing companies, McCauley and Hartzell are located within a few miles of each other in the same city of Dayton, Ohio, the Wright brothers' home town! It was on the request of Orville Wright to his friend Robert Hartzell to make propellers that saw the start-up of the Hartzell Propeller Company.

The First Turboprops

It is now common knowledge the Vickers Viscount V.630 was the World's first turboprop transport, which first flew on 29 July 1948. But, what was the World's first turboprop aircraft to fly? In 1930, Frank Whittle (1907–1996) (later Air Commodore, Sir Frank Whittle) patented his jet engine design with work beginning on the engine in 1937. However, the development of the turboprop engine dates back to 1925 through the work of a UK scientist Dr. Alan A. Griffiths (1895–1963) at the Royal Aircraft Establishment, Farnborough. His axial flow compressor and turbine, driving a propeller progressed as far as the wind tunnel testing stage in 1926; however, the research was brought to an end due to the 'depression' and lack of interest by the industry. It was to be another nineteen years before the turboprop idea was revised, following on after the development of the jet engine during the years of World War Two.

So, to return to the question of what was the World's first turboprop aircraft to fly. The aircraft was a converted Gloster Meteor F.1., jet fighter which first flew as a twin-engine turboprop aircraft on 20 September 1945. The Meteor's twin 750 SHP (559 kW) Rolls Royce Derwent 2 turbojet engines were modified with reduction gear and drive shafts to turn 7 feet 11 inch (2.14 m) five-blade Rotol props. The props were relatively small to allow for clearance between the tips and the fuselage. With this conversion installed, the engines were known as Rolls Royce R.B. 50 Trents. [It was Rolls Royce policy to name their engines after British rivers].

The Vickers Viscount 630 was the World's first transport turboprop aircraft, but it was not the first revenue-earning turboprop. This distinction was claimed by two converted Douglas DC-3 cargo planes, powered by Rolls Royce Dart R. Da 3/505 fixed-shaft turboprop engines. The two aircraft were placed in service by British European Airways in 1951–3 to become the World's first revenue-earning turboprop aircraft, with the first revenue-earning flight

The Gloster Meteor F9-40 prototype, similar to the Meteor that flight-tested the first turboprop engines. This example is located in the RAF Museum Cosford, UK.

1 – The History of Aircraft Propellers

The Vickers Viscount was the World's first turboprop airliner. This example, a 744 model, is located in the Pima Air & Space Museum, Tucson, AR.

from London to Hanover, Germany on 15 August 1951. The engines, more powerful than the prototype Viscount's original engines, were on flight trial in preparation for use on the Viscount Series 500 production aircraft. The Viscount's inaugural revenue-earning flight occurred on 18 April 1953 from London to Cyprus. Other Douglas DC-3's have also been converted to turboprop configuration over the following years.

Supersonic Propellers

High propeller tip speeds result in the separation of the air flow boundary layer over the blades, causing noise and a loss in efficiency due to compressibility problems. The problems of high tip speed were investigated as far back as 1949–58 in the USA and Europe, when research was conducted on propellers designed to operate in a supersonic air flow. Curtiss Electric built the first supersonic propeller to fly with the first test-flight taking place on the 14th April 1953. The four-blade, 10 feet (3.05 m) prop turned at supersonic speed powered by an Allison XT-38 turboprop engine mounted in the nose of a McDonnell XF-88B Voodoo prototype escort fighter (the forerunner of the F101). Research into supersonic propellers was continued at Edward Air Force Base, California, in 1955. A modified Republic Thunderstreak, the XF-84H, was used to flight-test the three-blade Aero Products supersonic propeller with a first flight on 22 July 1955. An Allison XT-40-A-1, 5850 ESHP turboprop powered the supersonic prop. The prop was adorned with a very large spinner to mate with the aircraft's nose, giving the propeller blades a short stubby appearance despite

The Republic Thunderstreak XF-84H used to flight-test the three-blade Aero Products supersonic propeller. The research aircraft is now housed in the National Museum of the USAF in Dayton, Ohio.

1 – The History of Aircraft Propellers

their 12 feet (3.64 m) diameter. The inaudible hypersonic sound waves emanating from the prop during ground running tests caused nausea to nearby personnel and combined with other problems with the prop and engine, the flight-test program was terminated. Many millions of dollars were spent on the project before supersonic props were deemed impractical due to their high noise levels and loss of efficiency at high operating speeds. The research was eventually brought to a close.

Record Breakers

The quest for high speed and improved aircraft performance has been present ever since the Wright brothers first flew. Speed records were being broken at a steady rate as improvements in aircraft, engines and propeller design allowed. The battle for the Schneider Trophy is a good example of the desire to fly faster. The trophy was captured by the British Supermarine S.6-b floatplane on 13 September 1931, flown by Flt. J. N. Boothman at an average speed of 340 MPH (295 knots). The Supermarine S.6-B went on to attain 407.5 MPH (354 knots) two weeks after winning the Schneider Trophy. The design experience gained on the S.6-B by R. J. Mitchell was to lead on to the design of the Spitfire of World War Two fame. The S.6-B used a 2300 BHP Rolls Royce R engine, which became the forerunner of the famed Merlin engine, which powered the Spitfire, Hurricane and Lancaster bombers.

The Italian Macchi MC-72 Castoldi, flown by Francesco Agello in 1931, was built to compete for the Schneider trophy but failed to enter. However, it did gain the World's speed record for floatplanes of 440.68 MPH (385 knots), which this author believes still stands in 2014. It is note

The Supermarine S.6-B Schneider Trophy winner on display in the Science Museum, London.

worthy the Macchi's engine was a Fiat 2850 BHP, which powered a contra-rotating propeller to combat prop torque; a very powerful engine for such a small plane! All speed records were held by floatplanes during the years 1928 through to 1939 due to the Schneider Trophy challenge, which was only open to floatplanes. The Macchi MC-72 held the overall world speed record for all classes of aircraft until 26 April 1939 when the German Messerschmitt Bf 109R raised the world speed record to 468.9 MPH (408 knots).

Other speed records of interest followed, taken by the Chance Vought F4-U Corsair, Grumman Bearcat and the TU 95/142. The F4-U Corsair fighter of World War Two fame had the largest propeller found on a fighter, radius 13 feet 2 inches (4.01 m). The prop was driven by the P&W R-2800 Double Wasp of 2250 BHP, enabling the Corsair to be the

1 – THE HISTORY OF AIRCRAFT PROPELLERS

first US fighter to exceed 400 MPH (350 knots) during a subsequent test flight on 10 October 1940.

Lyle Shelton's Grumman F8F Bearcat *Rare Bear* holds the present world speed record for piston/prop aircraft. The record was gained on 21 August 1989 at 528.31 MPH (460 knots) at Las Vegas, Nevada. The record for the world's fastest turboprop goes to the Russian TU 95/142 at 545.076 MPH (473 knots) on 9 April 1960. It is claimed by some authorities, the Republic Thunderchief, mentioned earlier with a supersonic prop achieved a speed of 670 MPH (580 knots) during flight test program to become the fastest propeller driven aircraft. However, the National Museum of the USAF in Dayton where the Thunderchief resides, records a lower speed. Did it reach Mach 1? This would have depended on the ambient temperature at the time (the speed of sound varies with temperature) and also, is the claimed speed of 670 MPH correct? No doubt, all these speeds records will be broken again some day in the future.

One of many flying examples of the Chance Vought F4-U Corsair taxiing out for take-off.

Propulsors

Other types of propellers still under development since the 1970s are the Propfan and Propulsor. Unlike the Propfan, which is a relatively new idea, Propulsors have been around for many years. A Propulsor is simply a propeller mounted inside a shroud. Although it has several advantages over conventional un-shrouded props, it has never been greatly utilized by aircraft designers, because of its suitability only for low-speed operations.

The Propulsor can trace its origin as far back as 1910, to the Bertrand Monoplane of French design. It had a single-engine driving a tractor and a pusher propeller, one at each end of the shroud. This was followed in 1932 by the Italian built Stipa-Caproni, with a 120 BHP DeHavilland Gypsy III engine and Propeller, which were both mounted inside the shroud, which was a part of the fuselage. A more recent example of a Propulsor equipped aircraft appeared in 1996, in the form of the experimental Bell X-22A with four turbine engines driving separate tilt-vector propellers mounted inside shrouds. The US/German, VFW Rhein Flugzeubau/Grumman American joint venture Fanliner with two seats and a pusher Propulsor was another example. The Brooklands/Edgeley Optika from the UK, designed for the observation role, was showing good prospects for being a commercial success until ten of the twenty aircraft built was destroyed in a factory fire. Some could possibly be still flying. However, prior to this loss, the prototype's first flight was made on 14 December 1979. The five-blade fixed-pitch propeller is powered by a 200/210 BHP Lycoming engine, giving the aircraft a cruise speed of 57–108 knots. Due to the propulsor, it is said to be the world's quietest aircraft. Present day airships, such as those built by Airship Industries of the UK, have relatively low cruising speeds and are ideally suited to being propelled by the Propulsor type of engine.

Propfans

In 1975, Hamilton Standard introduced a proof of concept advanced turboprop engine designed for short to medium haul airliners. The new engine was designed to power a propeller of advanced design, with five to thirteen sweptback, scimitar shaped blades. In 1976, NASA's Lewis Research Center in Cleveland, Ohio, contracted to Hamilton Standard to jointly research and developed an advanced turboprop engine driving a propeller with eight titanium blades to be efficient at the aircraft's cruise speed of Mach 0.8. Titanium was chosen because metal blades are too thick and therefore inefficient at the transonic speeds (Mach 0.8 to 1.2) the propeller was designed to run at. This propeller became known as the Propfan, a term now generally used to describe such propellers.

NASA later joined forces with General Electric (manufacturers of jet engines) to produce a Propfan using composite blades for their strength and light weight. General Electric copyrighted the name of their Propfan an 'Unducted Fan (UDF). The first flight of the GE-36 UDF occurred on 20 August 1986 at Edwards Air Force Base, California, and testing ended in February 1987. The engine was attached to the right-hand engine mount of a Boeing 727-100 test plane and was the first Propfan to become airborne. The unducted, pusher Propfan had an eight-blade contra-rotating propeller of 11 feet (3.35 m) diameter.

General Electric also flight-tested a Propfan on a McDonnell Douglas MD-80 between May 1987 and March 1988. This Propfan also had eight scimitar shaped contra-rotating blades, driven by the GE turbine engine with an

The McDonnell Douglas MD-80 flight-testing the P &W Allison Model 578-DX Propfan.
Photo courtesy Hamilton Standard, Connecticut, USA.

un-geared, direct drive, unducted fan. This resulted in a greatly improved performance over a conventional jet engine. McDonnell Douglas named their Propfan a UHB (Ultra-high By-pass ratio engine) with scimitar blades. It had a by-pass ratio of 36:1, compared to 5:1 ratio of a Boeing 747's JT5D engines. After four years of research and development at a total cost of around $100 million, McDonnell Douglas cancelled the project due to the high cost of the engine.

Pratt & Whitney joined forces with Allison to jointly test the P&W Allison Model 578-DX 20,000 static-pound thrust turboprop with a Hamilton Standard six-blade contra-Propfan of 11 feet 7 inches (3.53 m). The first flight of the MD-80 with the geared Propfan occurred on 13 April 1989, with all testing completed in a few weeks.

The extent of this author's research shows the big American companies are no longer involved in any Propfan research for several years now. In the Commonwealth of Independent States (formerly Russia) research continued for a longer period. An Ilyushin IL-76 test-bed aircraft was used to test the Lotarev contra-rotating tractor Propfan mounted on its number 2, (port inner) engine mount, with the first flight being made on 25 March 1971. The Yak-46, a 168-seat airliner with a pusher Propfan, is powered by two 24,690 pounds thrust ZMDB Progress D-27 Propfan with a first flight conducted circa 1995. The Antonov AN-180 was another type of passenger aircraft on the drawing boards around that time. It was due to fly around 1995 and was powered by two, rear-mounted tractor Propfan engines, the same type as found on the YAK-46.

Other types under test include the Antonov AN-70T designed as a military and civilian large STOL transport with a maximum all-up weight of 286,600 pounds (130,000 kg). It made a first flight on 16 December 1994 but its life was short-lived, due to its loss in a mid-air collision with its Antonov AN-22 chase plane on 10 February 1995. The second prototype first flew on 24 April 1997 with a third aircraft built in 1998. A 14,000 SHP Progress/Motor Sich D-27 turboprop engine drives each of the four, Aerosylva Stupino SV-27 turbine engines drive tractor, contra-rotating Propfans, which are different to the usual Propfan configuration. The fan has eight blades mounted on the front prop and six-blades mounted on the rear prop. Production was planned for this aircraft with 100 units being built for the Ukraine Air Force and 500 for the Russian Air Force, as of 1998. After a prolonged test program lasting sixteen years with many obstacles in its path, production started in 2012 for the Antonov AN-70. This places Russia ahead of the USA in Propfan technology, and with Propfan aircraft in service.

Conclusion

The propeller's development has come a long way since Blanchard first used a propeller on his hot air balloon way back in 1784, to the present day high-technology propellers. Constant-speed props with auto feathering and reverse thrust have been around for several years now and composite props are being used more and more on new aircraft types. Will the Propfan replace the present day turboprop transport aircraft? It may do so one day in Russia, but for the rest of the western world, turboprop aircraft with at least six-blade composite props will be flying for many years to come.

A wide chord prop with squared off tips adorns this Yak-55M aerobatic aircraft.

2 – Propeller Pitch

The Purpose of the Propeller

The purpose of the propeller is to convert the engine torque into axial thrust, or propwash. To provide the necessary force to propel the aircraft forwards, the prop displaces a large volume of air rearwards. Newton's third law is obeyed by the equal and opposite reaction force of prop thrust acting in a rearward direction. How well the propeller achieves this is measured by the prop's efficiency. This is determined by a number of factors, which either improve or reduce the propeller's efficiency. These factors include the pitch, blade angle, diameter, solidity, number of blades, tip speed, drag and the location of the prop in relation to the engine's nacelle or the fuselage, chord variation along the blade, the shape of each blade element and the prop tips. Also included are the lift and drag coefficients, which are a function of the angle of attack of the propeller blades.

All these factors affect the propeller's absorption of engine power and its ability to convert the propwash (working fluid) into thrust and will be considered in turn throughout this book, followed by the forces acting on the propeller.

The Working Fluid

The volume of air in which the propeller works is the working fluid, and is described as an inviscid fluid flow. The term inviscid describes the air flow and assumes it to be one without viscosity, or devoid of any internal friction or drag force. The airflow approaching the prop is termed the relative air flow (RAF) and shown by three arrow heads on the diagrams; it is also known as the inflow velocity. The air flow behind the prop is termed the propwash or outflow.

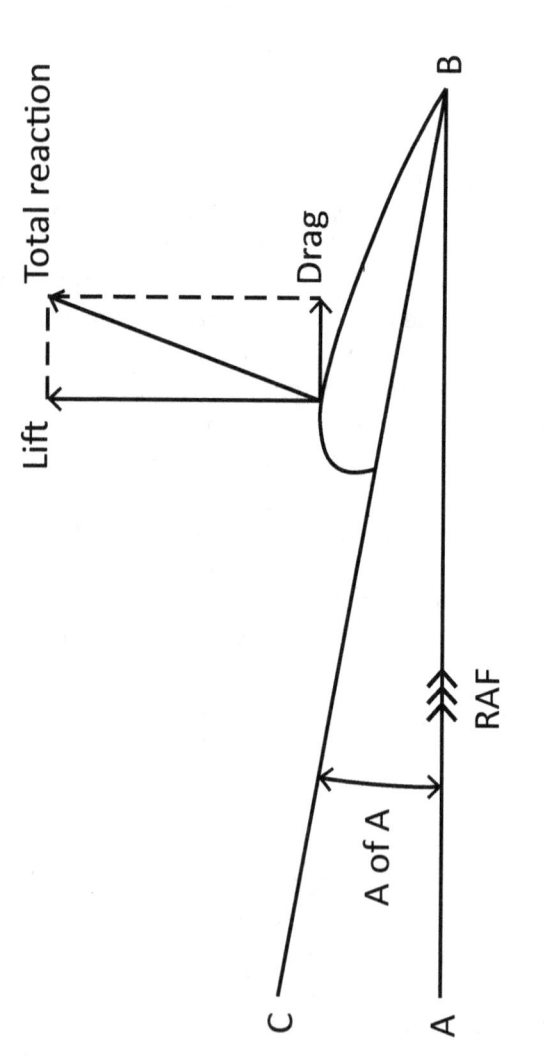

Diagram 1, Airfoil Terminology

Propeller Terminology

All pilots should be familiar with the diagram opposite, Diagram 1, Airfoil Terminology, from their previous study of classical aerodynamics. The diagram depicts the terminology used in defining the wing's airfoil section, chord, relative airflow, angle of attack, lift, drag and total reaction.

Briefly, the line A–B represents the path the airfoil is moving along. The relative airflow is therefore moving along the same path, but in the opposite direction and depicted by the three arrow heads. The line C–B represents the extended chord line and the direction the airfoil is pointing. The angle between A–B and C–B, that is, between the relative airflow and the chord line, is the angle of attack. Due to the angle of attack, the air mass flowing over the airfoil will produce an aerodynamic reaction known as the total reaction (TR), which can be divided into the forces of lift and drag. The propeller is a rotating wing and shares the same terminology as the aircraft's wing. Because the prop is rotating around its axis combined with the forward motion of the aircraft, it requires additional vectors and terminology to fully describe it, as follows:

- The chord line is taken to be a tangent to the lower surface of the prop blade.

- The blade element is a theoretically thin cross section of the prop blade (it is the counterpart of the wing's airfoil section) and is perpendicular to the blade's major axis. The blade profile is the shape of this cross section of the blade.

- Confusing as it may sound, the propeller's back is the curved, upper surface and this part of the prop is viewed from in front of the aircraft.

2 – Propeller Pitch

- The prop blade's relatively flat surface corresponds to the wing's under surface and is known as the prop blade's face, thrust face or pressure face due to this side of the propeller producing air pressure above ambient, but more of this later in 'Propwash Thrust'. The face is usually painted a matt black to reduce reflection making it easier for the pilot to see through the prop disc in flight.

- The shank is located at the blade root and being circular; it is not an aerodynamic shape and therefore plays no part in producing thrust, although it does produce some drag. Not all propellers have shanks.

- The propeller's boss is the thick central non-aerodynamic part of a wooden fixed-pitch prop.

- The hub is in the same central position as the boss but is a separate unit to which the blades and constant-speed unit are attached.

- The spinner is the streamline fairing covering the hub area. On some aircraft it is used for aesthetic reasons or it may also be an essential item on other aircraft to smooth the airflow into the engine air intakes, and reduce drag.

- The prop blade's leading and trailing edges, plus the tips and roots, all share the same terminology as applied to an aircraft wing.

- The blade cuff is located at the blade root to enhance air flow into the engine intakes and for greater propeller solidity.

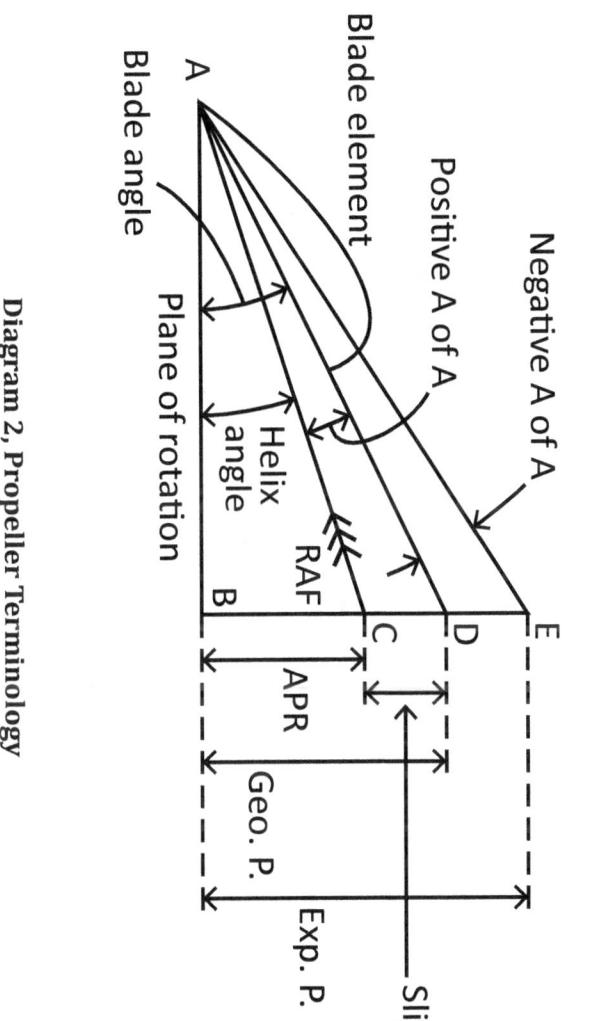

Diagram 2, Propeller Terminology

2 – Propeller Pitch

Comparing Diagram 2, Propeller Terminology with Diagram 1, it is shown the blade element, chord line, relative airflow and angle of attack are all shown similarly, albeit reversed. [Classical aerodynamic diagrams by convention are shown as moving from right to left, and the reverse direction for propeller aerodynamic diagrams]. Diagram 2, also shows the relationship of the helix angle AB–AC, the blade angle AB–AD, the angle of attack (AC–AD), all measured in degrees. The advance per rev (B–C), the geometric pitch (B–D), the slip (C–D) and the experimental pitch (B–E) are all measured in inches; vector B–E represents the prop's axial (or forward) component. Using a breakdown of Diagram 2, each factor can now be examined in more detail, keeping the following points in mind. All that follows will also apply to a constant-speed propeller; however, it will be assumed for now a given blade section on a fixed-pitch propeller is being considered where the geometric pitch and experimental pitch both remain constant and the advance per rev and slip are variables. Also, vector A–B represents the tangential component or plane of prop rotation.

Propeller Pitch

What is the definition of pitch? And what is the difference between the experimental pitch, the geometric pitch and the advance per rev? The difference will be revealed as each term is considered in turn.

The definition of the term 'pitch' and its analogy with the common screw must be defined. When a screw is screwed into a solid medium such as wood, it will advance a given distance in each turn equal to its pitch; in other words, its advance per revolution is a fixed quantity equal to its pitch.

The geometric pitch of the screw is the distance between each adjacent thread.

When an aircraft is gliding with the engine/prop stopped, the propeller's advance per rev is infinite, as opposed to a stationary aircraft with the engine running and the prop's advance per rev is zero. The advance per rev of the propeller is a variable quantity depending on the aircraft's speed, and propeller RPM, or lack of either, between these two extremes.

Some text books state the analogy with the wood screw is useless, because the screws advance per rev is a fixed quantity, while the props advance per rev is a variable quantity between zero and infinity. That is true, but there has to be a condition of propeller operation somewhere between these two extremes that satisfies the definition of geometric pitch. And there is! When the aircraft reaches a given forward speed, the propeller blade's angle of attack will become zero and the advance per rev will equal the geometric pitch, which is exactly what happens with a wood screw.

In conclusion, two things have been proven, one, the screw analogy is a true and useful example to introduce the definition of pitch, and two, under certain operating conditions of aircraft forward speed and prop RPM, the advance pre rev equals the geometric pitch.

Helix Angle

The helix angle (AB–AC) which is also known as the angle of advance, is the angle between the propeller's plane of rotation (A–B) and the resultant direction of the relative airflow (RAF), vector A–C in Diagram 2.

When the engine is running the prop will have a rotational velocity, vector (A–B) in Diagram 2, and will

2 – Propeller Pitch

travel on a circumferential distance equal to $2\pi R$ in unit time in the plane of rotation. The rotational velocity is also known as the tangential velocity. When the aircraft is moving forward, it will have a forward velocity along the axial vector component (B–C) and it will cover a given forward distance known as the 'advance per rev' (or the effective pitch) in unit time depending on the forward speed of the aircraft. Any change in prop RPM or advance per rev will induce a change in the helix angle (AB–AC). The helical flight path followed by the chosen blade element will be along the vector A–C. The helix angle is related to the advance per rev, or effective pitch (B–C).

The helix angle can be found from the following formula:

$$\text{Helix angle, } \tan \theta = \frac{P}{2\pi R}$$

Where P = B–C axial component or effective pitch (APR)
R = prop radius
π = 3.14...

Therefore, the effective pitch can be found from the formula:

$$\text{Effective pitch (P)} = 2\pi R \tan \theta$$

The propeller tip's helical path will be approximately 45° from the vertical and increases towards the blade root. The blade tip helix angle will also vary from zero degrees when the aircraft is stationary through approximately 45° at the design cruise speed, to a greater angle as the aircraft's speed increases above its design cruise speed.

Blade Angle & Twist

The blade angle is defined as the angle between the propeller's plane of rotation (A–B) on Diagram 2, Propeller Terminology, and the prop blade's chord line (A–D) combing the helix angle plus the angle of attack. It has the same meaning as a wing's angle of attack.

Each blade element travels on a different helical path and to produce the maximum lift/drag ratio must meet the relative airflow at the same small angle of attack of 3–4°. To achieve this constant angle of attack, along the length of the blade, the propeller blade must be twisted. This is known as the propeller's geometric twist where the angle between the blade chord and its plane of rotation varies along the blade's length. This requires the blade angle to be greater at the root with a gradual reduction towards the tip, as mentioned above. The geometric pitch of the propeller then remains constant (geometric pitch = 2πR tan θ) due to the blade angle decreasing with an increase in blade radius. The actual blade twist is designed to provide the correct angle of attack at the design cruise speed.

Although the blade twist is associated with the geometric pitch, it must not be confused with the definition for blade angle and pitch. The blade angle is measured in degrees between the vectors A–B and A–D, while the pitch is measured as a length in inches (or centimetres) along the vector B–D.

Experimental Pitch

The experimental pitch can be defined as the 'prop's advance per revolution when producing zero net thrust'.

An inspection of Diagram 3, Experimental Pitch, shows as the advance per rev (APR) increases from B to E, the

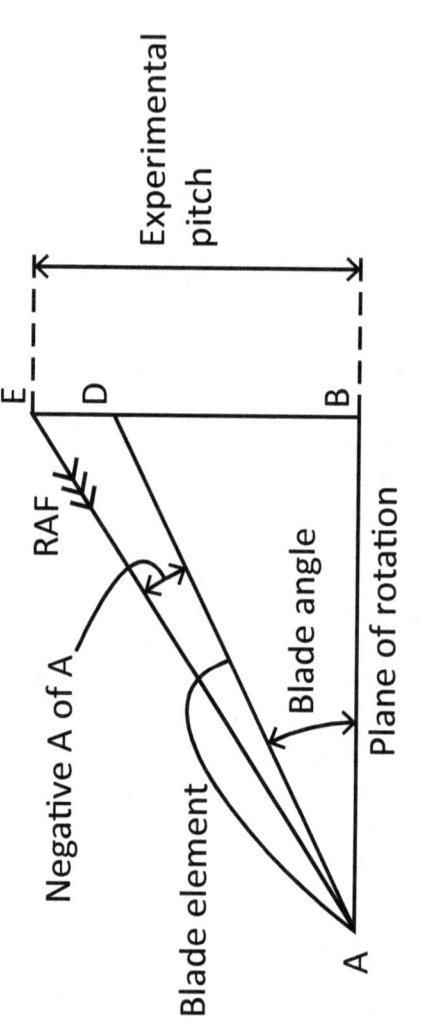

Diagram 3, Experimental Pitch

angle of attack will reduce to a negative angle (AD–AE) of around minus two degrees and the propeller will cease to produce thrust. In this condition the relative airflow acts along the line from E to A, and corresponds to the wing's 'zero lift line', to become the prop's zero thrust line. This is the important aerodynamic feature of the experimental pitch. From the designer's point of view, it is considered as being the 'ideal pitch'. Because it has a definite value and length, depending on the prop's characteristics, it can be used for experimental measurements, hence the name.

A fixed-pitch propeller has only one experimental pitch, while a constant-speed prop's pitch is variable over the available operating range of the blade angles between the fine and coarse pitch stops. The experimental pitch may also be known as the 'zero thrust pitch' or the 'exponential mean pitch'.

Geometric Pitch

The geometric pitch is defined as 'the distance the propeller advances forward in one revolution when the angle of attack of the blades is at zero degrees'.

Diagram 4, Geometric Pitch explains this clearly. When the prop is advancing with zero degrees angle of attack, the advance per rev is equal to the geometric pitch. This distance is a definite length measured in inches for a given propeller (as mentioned above) which depends on the geometry of the blades, hence the name geometric pitch. To maintain a constant pitch, the angle of each blade section must increase from the blade tip to the blade root in order to obey the law, geometric pitch = $2\pi R \tan \theta$. This was explained earlier in the section on Blade Angle. The equation should hold true for the whole length of the prop blade, but if it does not, the geometric pitch is stated for

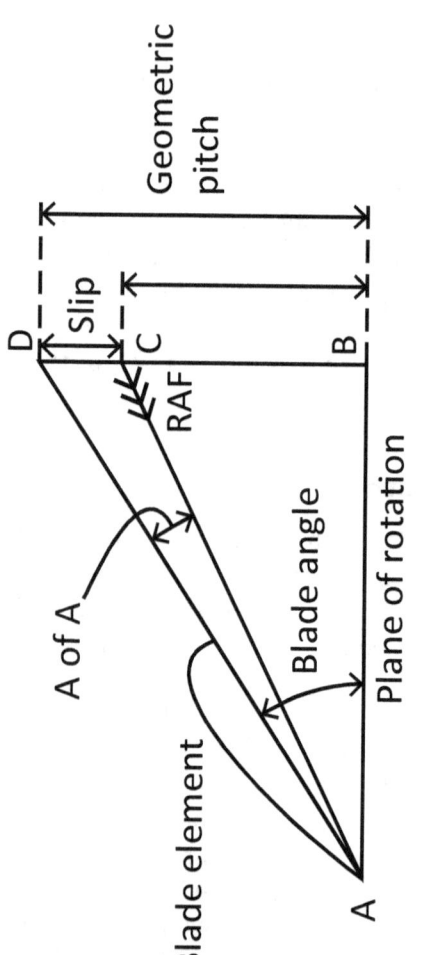

Diagram 4, Geometric Pitch

one section of the blade only at the 'standard radius; this point is located at the 75% station along the length of the prop blade from the hub. From the above formula, we can find the geometric pitch of a propeller in inches as follows:

Geometric pitch = $2\pi R \tan \theta$

Given: Pitch = 22° (tan 0.4040...)
   Propeller radius = 26 inches
   π = 3.14...

Therefore pitch = 2 × 3.14... × 26 × 0.4040...
   = 66 inches

An inspection of Diagram 4, Geometric Pitch, will reveal a small amount of slip is still present. A small amount of thrust is still generated by the prop at zero degrees angle of attack, which can be attributed to the curved shape of the propeller's back. Generally, the geometric pitch is less than the experimental pitch; however, this may not always be true.

To clarify the points made here regarding the experimental and geometric pitch, the prop blades can be related to the aircraft's wings and tailplane. The vertical tail is a symmetrical airfoil and so both sides are of equal curvature. However, on most aircraft (aerobatic aircraft can be an exception) the main wing is cambered; the upper surface has greater curvature than the lower surface. The propeller blades are shaped similarly. An inspection of Diagram 5, Lift Coefficient V. Angle of Attack, shows the symmetrical airfoil section ceases to produce lift at zero degrees angle of attack. However, of greater interest here, at zero degrees angle of attack, the cambered airfoil is still producing lift, or thrust in the case of the propeller as indicated by a positive lift coefficient. This point corresponds to the prop's geometric pitch. A further

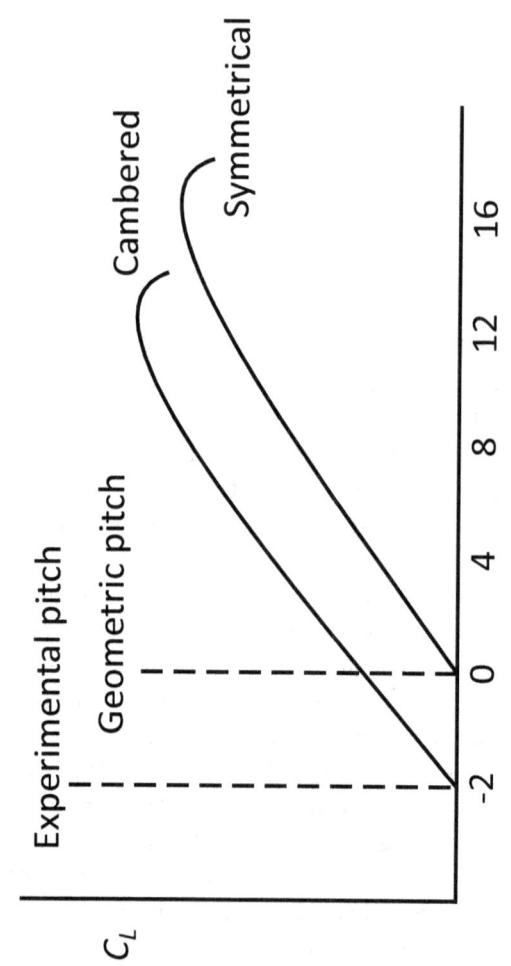

Diagram 5, Lift Coefficient V. Angle of Attack

reduction in angle of attack to minus two or three degrees will eventually produce zero lift coefficient, known as the angle of zero lift for the wing or the angle of zero thrust for the propeller. This corresponds to the prop's experimental pitch.

Slip

The 'slip' of the propeller can be defined as the 'difference between the advance per rev and the geometric pitch'.

When the advance per rev equals the experimental pitch, the angle of attack is slightly negative and produces zero thrust. However, in normal operating conditions, the angle of attack is around positive three degrees and the advance per rev is considerably less than the geometric pitch. Diagram 4, Geometric Pitch, shows the advance per rev (B–C) plus the slip (C–D) is equal to the geometric pitch (B–D). For the propeller to provide maximum thrust and efficiency, and because air is not a solid medium, slip must be present. Maximum efficiency is only obtained when the slip, expressed as a percentage is around 30% of the length of the geometric pitch. In other words, slip is the difference between the actual distance the prop travels forward in one revolution (B–C or effective pitch) and the distance the prop would theoretically travel in on revolution if its advance per rev were equal to the geometric pitch (B–D). Slip is a percentage of distance.

Given the prop RPM, prop pitch in inches and true air speed in knots, the slip can be found using the following formula:

$$\text{Slip} = \frac{\text{RPM} \times \text{pitch} \times 60}{6080 \times 12}$$

Given: RPM = 2400
Pitch = 69 inches
TAS = 100 knots

$$\text{Then slip} = \frac{2400 \times 69 \times 60}{6080 \times 12}$$
$$= 136.18 \text{ inches} = 26.57\%$$

It must be emphasized that slip is related to the geometric pitch and the advance per rev. As stated above, zero thrust occurs at a negative angle of attack (experimental pitch) and thrust increases as the angle of attack increases through the geometric pitch up to some positive angle of attack. Diagram 4, Geometric Pitch, shows a small amount of slip (C–D) is present when the prop blades are operating at zero angle of attack (geometric pitch). To produce thrust, slip must be present with the maximum prop efficiency occurring at around 30% slip, or 30% of the geometric pitch. It must be remembered slip is related to geometric pitch, which is the distance the prop travels during each revolution.

Brief Review

Before moving onto the next section, we will briefly review the differences between the slip, advance per rev, experimental pitch and the geometric pitch.

- Slip is a variable and is the difference between the advance per rev and the geometric pitch.

- The advance per rev (effective pitch) is less than the geometric pitch and experimental pitch under normal operating conditions. The advance per rev is also a variable.

- When the advance per rev is equal to the experimental pitch, the angle of attack is slightly negative with zero slip and thrust.

- When the advance per rev is equal to the geometric pitch, the blade's angle of attack is zero with a small amount of slip and thrust still present.

- The experimental pitch is a constant.

- The geometric pitch, which is also a constant, is usually less than the experimental pitch.

- On diagrams 2, 3 and 4, note the location of the relative air flow (RAF) vector.

In conclusion and referring back to Diagram 2, Propeller Terminology, the following points should now be obvious. All sections of the propeller blade have an advance per rev equal to the aircraft's forward speed. The vector A–B, represents the rotational velocity of a given blade element. The vector A–C represents the resultant direction of motion of the chosen blade element. Because the length of the vector A–B varies with each individual blade element, it follows each blade element travels on a different helical flight path with its rotational velocity increasing from the hub to the tip.

The vector A–B increases with increasing RPM and therefore the effective pitch (advance per rev) will also increase vector B–C. Conversely, the effective pitch will also increase if the aircraft's forward speed is increased. Therefore, the effective pitch increases along with the helix angle, which in turn increases the prop efficiency and thrust up to a certain point. As the advance per rev increases, the blade's angle of attack decreases and reduces the thrust and efficiency. This is where a variable-

2 – Propeller Pitch

pitch or constant speed propeller becomes a necessity in order to increase the blade angle and geometric pitch to maintain the required angle of attack, for the increasing combination of prop RPM and the aircraft's forward speed, known as the speed ratio.

Advance/Diameter Ratio

The aircraft designer has to choose the propeller with the most suitable pitch and diameter for the aircraft and its intended mission and design air speed. When confronted with a family of props which have their blade angles increasing in some systematic order, one useful parameter to refer to is the advance/diameter ratio. The advance/diameter ratio can be used to define the characteristics of a prop using a non-dimensional form. The advance/diameter ratio (J) is the ratio of the aircraft's velocity (TAS) to the product of propeller RPM and diameter.

$$\text{The advance/diameter ratio, } J = \frac{V}{ND}$$

Where, V = true air speed
       N = RPM
       D = propeller diameter

Diagram 6, Advance/diameter Ratio shows the curve for the advance/diameter ratio plotted against efficiency for a family of props with their pitch increasing. The numbers above the curves shows the blade angles for each propeller. The efficiency increases with an increase in the ratio up to a certain limit. At too high a ratio, the angle of attack of the blades exceeds the stalling angle at low forward speeds reducing the thrust available for take-off. Reducing the propeller's diameter also reduces the efficiency by over

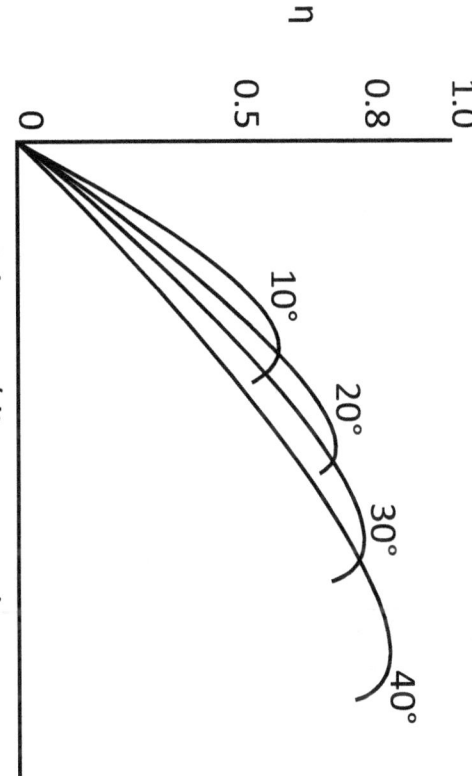

Diagram 6, Advance/diameter Ratio

2 – Propeller Pitch

working the prop blades, but more of this later. A fixed-pitch propeller's efficiency curve moves to the left with a decrease in RPM and TAS and to the right with an increase of RPM and TAS power. Diagram 6, also shows the curves for a fine/flat pitch prop and a coarse pitch prop, 10 and 40 degrees respectively. Most single-engine light aircraft have a prop between these two extremes. A constant speed propeller has an infinite number of curves between the fine and coarse pitch limit stops; the pilot selects the optimum setting for climb or cruise.

Slip function and effective pitch are alternate names for the advance/diameter ratio.

Fixed-pitch Propellers

It has been assumed up to now, the propeller to be of the fixed-pitch type, as found on low performance aircraft. The advantages of the fixed-pitch prop are its simplicity of operation for low time pilots, it is the cheapest type to install on an aircraft and it is relatively maintenance free due to the absence of a constant-speed unit (CSU). Its disadvantage is it gives its maximum efficiency at only one air speed, known as the 'design air speed'. At any other speed, above or below the design air speed, the prop efficiency will be reduced. However, in general, a prop will normally be chosen from a family of props to suit the aircraft's design air speed. In order to produce the maximum amount of thrust for the least amount of drag or torque, each blade element is set to a different angle to ensure the optimum angle of attack is maintained at the prop's speed ratio to produce its maximum efficiency. This is the reason why the prop blade is twisted. If the blades were not twisted, the blade root would be operating at a

FIXED-PITCH PROPELLERS

This beautifully restored Avro Anson Mark 1, of WW II vintage has fixed-pitch, two-blade, wooden propellers powered by two Armstrong Cheetah IX seven-cylinder radial engines of 350 BHP each.

negative angle of attack while the prop tip would be stalled when operating at the design speed ratio.

At the design air speed, maximum efficiency also depends on the geometric pitch. A prop with a relatively short geometric pitch will give the aircraft a better rate of climb over a prop with a longer geometric pitch. The short pitch prop is more suitable for a training aircraft or aerobatic aircraft which spend a relatively greater proportion of their flying time at lower air speeds, doing training manoeuvre and climbing to altitude. Conversely, a longer pitch prop produces a slightly higher cruising speed, favouring an aircraft used for cross-country flying. Longer pitch props may not attain full RPM at the start of the take-off roll, due to the blade's high angle of attack causing too much blade drag. However, if the pitch is too fine, the RPM will reach a maximum with the aircraft

2 – PROPELLER PITCH

stationary or early in the take-off run. On reaching cruising speed the engine RPM would exceed its limit, calling for a reduction in engine power. The American FAA certification rules require the propeller pitch to be such that it prevents the engine from over-speeding at maximum RPM, while climbing at the 'best rate of climb speed'. Likewise, the engine's RPM is not allowed to be exceeded by more than 10% in a dive at the never exceed speed (VNE) with the throttle closed. The same rules apply to constant-speed propellers.

It is now obvious, selecting a propeller with the desired pitch is very important. Propeller manufacturers list a selection (or family) of props designed for certain engines to aid the aircraft designer in his/her choice of propellers. There can be found on the prop hub a set of numbers such as 72" × 57". The first number refers to the diameter of the prop in inches, while the second number refers to the geometric pitch in inches at the 'standard radius'.

When flying a plane with a fixed pitch prop in conditions of turbulence, you may notice some rapid variation in RPM. This could be very disconcerting, inducing you to suspect engine trouble. The cause of the engine RPM variations can be attributed to the propeller loading and un-loading. As the aircraft attitude is constantly changing in the turbulence, the air flow through the prop disc will meet the blades at varying angles of attack causing variations in prop loading and hence a change in thrust and RPM. The throttle should be set to maintain the required RPM for turbulence penetration air speed (Vb) and be left there. The RPM will fluctuate around the desired setting, so do not chase it with the throttle; set it and forget it!

Variable Pitch & Constant Speed Propellers

Back in the early days of aviation, the limitations of the fixed pitch prop soon became evident with the advent of higher-powered engines and the greater speed range of newer types of aircraft. Since the 1930s, high performance aircraft have used either a variable pitch or more commonly, constant-speed props.

Pilots quite often incorrectly refer to constant speed props as variable pitch props. It is true, the constant speed propeller does have a variable pitch change mechanism, but there is a difference here. The variable pitch props are simply that – variable; that is, they do not have constant speed ability. Variable pitch props can be classified under three basic headings of ground adjustable, two position, or controllable props. Two different methods are employed for changing the propeller's pitch. On the ground adjustable type, after loosening the collar bolts on the round shank at the blade root, the blade angle is then adjusted to the

The DeHavilland Mark 1 Heron was the world's smallest airliner with four prop/engines.

2 – PROPELLER PITCH

required pitch; the bolts are then re-tightened on the collar. The second method is more convenient for the pilot, the pitch being adjusted by a control in the cockpit while in flight. The fine (or flat) pitch position is selected for take-off and climb. On reaching cruise altitude, the pitch control lever is moved to select coarse pitch. The controllable pitch propeller works on the same principle as the two-position prop but with the addition of a number of intermediate pitch positions available for selection between the fine and coarse pitch stops. Because a given pitch is selected at any one time, the engine RPM will vary in the same manner as a fixed pitch prop with changing air speed, power settings and prop loading. It is important to remember, a VP prop does not have a constant speed unit (CSU) and therefore will not maintain a constant RPM. The types of VP props

The Lockheed (L-1049) Constellation was the last of the piston/prop airliners to be built. It has four Wright R-3350 radial engines of 2200 BHP each, driving Hamilton Standard or Curtiss Electric constant-speed propellers. This aircraft is located in the Pima Air & Space Museum, Tucson, Arizona.

mentioned above are the most common types but, there have also been other less popular, or should I say, novel types?

In 1934, a D.H. 88 Comet won the London to Melbourne air race. The two, 230 BHP engines powered two-blade VP French Ratier props. The Ratier's unusual feature was their method of changing the blade angle. This was achieved by pressurising the pitch change mechanism cylinder with a bicycle pump to turn the blades to fine/flat pitch for take-off and climb. As the air speed increased, dynamic air pressure acting on a disc on the front of the prop spinner overcame the cylinder's internal compressed air pressure and turned the blades to coarse pitch; and there they stayed for the remainder of the flight. There was no way to alter the pitch while airborne and the landing (and go-around if necessary) was flown with coarse pitch. After landing, the cylinder was re-charged ready for the next flight.

There are a few aircraft in production today using variable pitch propellers. For this author, the Denny Kitfox and the Burkhart G109 Grob motor glider are two names that come to mind, however there are several homebuilt and ultralight aircraft that use VP propellers. The original Kitfox used a three-blade, wooden, ground adjustable variable pitch propeller. The Grob has a two-blade, VP prop with a choice of three different pitch settings of fine, coarse and feathered. The feathered position is selected after the engine is shut down at altitude and the aircraft is flown as a glider, with the fine and coarse pitch settings being used in the normal manner for take-off and climb.

The idea of having a selection of pitch settings on the VP prop was soon refined to produce the constant speed prop so widely used today, made possible by the invention of the constant speed unit (CSU) located at the prop hub,

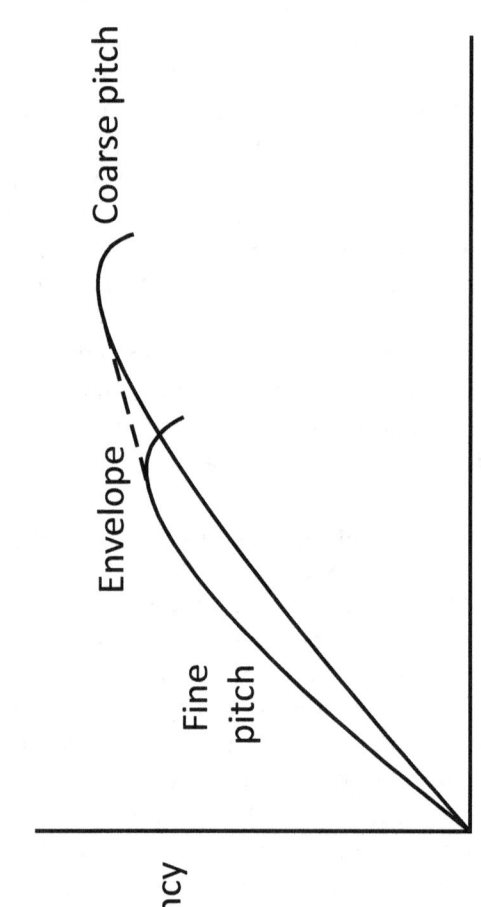

Diagram 7, VP & Constant Speed Props

VARIABLE PITCH & CONSTANT SPEED PROPELLERS

The Beech UC-45J Navigator has a pair of Hamilton Standard variable-pitch props each powered by a P&W Wasp 450 BHP engine. The Pima Air & Space Museum in Tucson, Arizona is home to this aircraft.

which may or may not be covered by a prop spinner. The CSU will be covered in greater detail later in this book.

A diagram for a VP prop would show just two curves, for fine and coarse pitch only, as opposed to a diagram for a constant speed prop, which has an infinite number of curves. Refer to Diagram 7, VP & Constant Speed Props, which shows the fine and coarse pitch performance curves for a VP propeller, which also represent the fine and coarse pitch limits for the constant speed prop with an infinite number of performance curves for the constant speed envelope. The dashed line along the top of the curves indicates the infinite performance curves and efficiency over the constant speed range. Apart from an overall increase in efficiency, the constant speed propeller has several other advantages over the fixed-pitch prop. Obviously, the RPM remains constant, hence

8A, Fine pitch

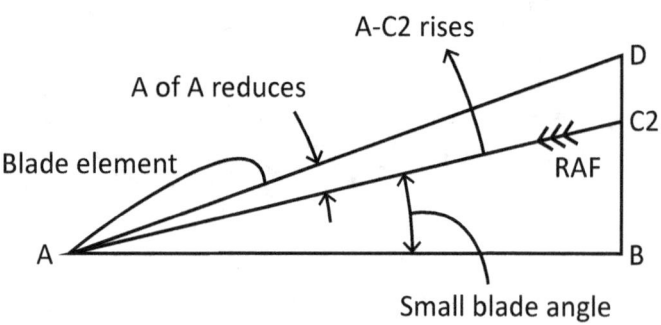

8B, Coarse pitch

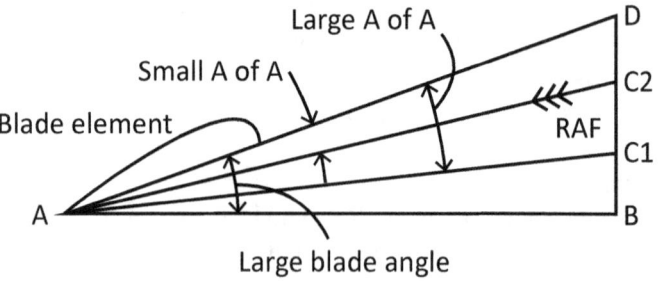

Diagram 8, Fine & Coarse Pitch

Variable Pitch & Constant Speed Propellers

the name, along with constant power for a given manifold/RPM setting. Also, decreasing air density with altitude is compensated for by an automatic increase in propeller pitch by the CSU.

Diagram 8, Fine & Coarse Pitch, shows the advantages and disadvantages for using fine and coarse pitch for take-off and cruising respectively. Diagram 8A, shows the condition with fine/flat pitch selected for take-off, the usual setting. The small blade angle AB–AD and the resulting small angle of attack (AC2–AD) gives the thrust and efficiency required to provide the maximum acceleration to reduce the take-off run and to produce the maximum rate of climb. As the aircraft accelerates to cruise speed after levelling off at cruise altitude, the relative air flow line (A–C2) will 'rise' towards the extended chord line (A–D) as the advance per rev increases, in effect, increasing the length of the advance per rev line (B–C2). This results in a reduction in the angle of attack and therefore, a reduction in thrust and efficiency. This proves the necessity to increase the prop pitch to a coarse setting for cruise flight.

In Diagram 8B, Fine & Coarse Pitch, coarse pitch has been selected for take-off, which should never be used, by choice. The large blade angle (AB–AD) with its associated large angle of attack (AC1–AD) could possibly result in stalled propeller blades giving considerably reduced thrust and poor acceleration on take-off. Assume now, the correct setting of coarse pitch has been selected for the cruising speed. Due to the large blade angle (AB–AD) the advance per rev (B–C2) and the reduced angle of attack (AC2–AD) will be at the optimum angle of attack of about three degrees providing good efficiency.

The prop control is set to full fine/flat pitch for take-off and landing but for the remainder of the flight, the pilot selects a chosen engine RPM, not pitch. The

2 – PROPELLER PITCH

A DeHavilland DH.98 Mosquito on take-off
with both props in fine/flat pitch. This is the
World's only airworthy Mosquito in 2014.

constant speed unit to maintain a constant engine RPM is constantly adjusting the prop blade angle. If the manifold pressure is increased, the CSU will increase the blade angle automatically to absorb more engine power without any increase in RPM. Even with full fine/flat pitch set for take-off, the CSU will turn the blades to a slightly coarser pitch setting to prevent over speeding as the prop load is reduced with increasing take-off speed.

The prop will only 'constant speed' between its fine and coarse pitch limit stops. Below a pre-determined RPM, usually around 1500–1600 RPM on light aircraft piston-engines, the prop blades will reach their fine/flat pitch limit stop and RPM will vary the same as on a fixed pitch propeller with changing aerodynamic loads, air speed and power settings. This will occur for example during the approach to land with the throttle partly closed. The

Variable Pitch & Constant Speed Propellers

fine/flat pitch limit stop provides the optimum angle of attack for low speed operations, such as during take-off and landing. Some turboprop aircraft have a 'ground fine/flat pitch' setting; this is an ultra-fine/flat pitch setting with an angle of attack less than fine/flat pitch for ground operations only. It produces zero thrust while taxiing and saves on brake wear. The additional blade drag will reduce the landing roll distance for an aborted take-off. The use of ground fine/flat pitch is known as 'discing'.

If the aircraft is placed into a steep descent, the constant speed unit will turn the blades towards coarse pitch to maintain the selected RPM. However, once the blades reach the coarse pitch limit stop, the RPM will increase along with increasing air speed; the prop is driven partly by the force of the air flow through the prop disc. The coarse pitch stop is there to prevent the prop blades moving in to an over-coarse pitch setting and prevent the propeller from over-speeding. When 'feathering' the propeller, the coarse pitch stop is removed, but more on feathering shortly.

The variable pitch range of a constant speed propeller will spread the design speed over a greater range of speeds, as opposed to only one air speed for a fixed pitch propeller. For either the fixed pitch or constant speed prop, the blade twist can only be suitable for one speed – the design air speed. At any other speed, off-point losses will occur but will be less for the constant speed propeller.

Diagram 9, Prop Load Curve, clarifies the point made above and in the previous section on fixed pitch propellers. Curve 'A' is the full throttle curve and represents the maximum power available from the engine at any given RPM. Curve 'C' represents the power absorbed by the propeller in fine/flat pitch; in this example, it produces the 200 BHP available at full revs (2700 RPM) indicated where the curves 'A' and 'C' coincide. Curve 'B' represents the

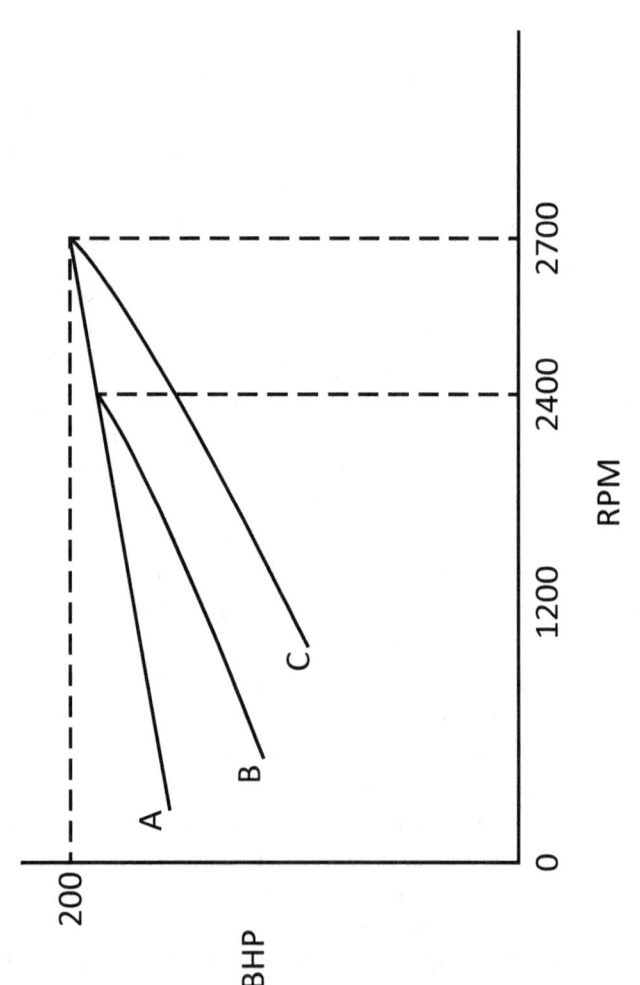

Diagram 9, Prop Load Curve

power absorbed by the prop in coarse pitch. Full throttle is achieved before maximum RPM or maximum BHP is reached. This is usually the situation for a fixed pitch prop; on opening the throttle fully for take-off, the RPM peaks at around 2400 RPM until the aircraft's forward speed increases and then the RPM will gradually increase to its maximum value. Some pilots refer to fine/flat pitch as flat pitch (common terminology in the USA). Same thing – different name.

Larger aircraft have the range of pitch change operation increased to include reverse thrust. The pitch range around the fine/flat pitch setting through reverse pitch is referred to as the 'Beta range'. The use of reverse pitch is known as Beta mode, and when selected, the normal operation of the constant speed unit is de-activated and the pilot has direct control over the propeller pitch (via the throttle levers) to control reverse thrust for landing roll braking. This will be covered in greater detail later in the section on Propeller Operation.

3 – Thrust & Efficiency

Efficiency

In the early part of this book under the section 'The Purpose of the Propeller', it was stated, "The purpose of the propeller is to convert the engine torque into axial thrust, or propwash". The statement can now be rephrased to read "the propeller produces the greatest axial thrust for the least amount of engine torque, when the maximum thrust/torque ratio is being produced". How well the propeller converts the engine torque into axial thrust is measured by the propeller efficiency which in turn depends on several factors, namely, the prop's diameter, solidity, number of blades, and prop blade loading to name a few. All of these factors and more will be covered in this chapter. Efficiency is therefore the best way to measure a prop's performance; however it is not the whole story as will be revealed later. Thrust is a force that propels the aircraft through the air, but the efficiency is a measure of how well the prop succeeds in achieving this objective.

Given figures for a fictitious light aircraft, the propeller's efficiency can be easily calculated using the following formula:

$$\text{Efficiency} = \frac{60 \times TV \times 100\%}{BHP \times 33{,}000}$$

Given: BHP = Brake Horsepower = 200
T = Thrust = 385 ponds
V = Air speed = 240 FPS = 142 kts

$$\text{Efficiency} = \frac{60 \times 385 \times 100\%}{200 \times 33{,}000} = 84\%$$

3 – Thrust & Efficiency

The above answer shows the prop's efficiency to be 84%, which is a fairly good result; most metal props have a peak efficiency of around 80%. The remaining 16% of engine power is used in counteracting the losses from friction, ancillary drive and exhaust gasses, etc. From the above formula a curve can be drawn with efficiency and true air speed as parameters, as in Diagram 10, Efficiency V. TAS. The curve is drawn for a fictitious aircraft example with a metal prop, showing the curve peaks at 84% prop efficiency at the design air speed of 142 knots. Above and below this figure the maximum efficiency deteriorates for a given speed. The lower curve is drawn for a theoretical wooden prop and shows its maximum efficiency peaks at 70%, due to greater blade thickness required for strength. A wood prop is not as structurally strong as a metal prop and so has to be built of thicker materials for extra strength, which is not as aerodynamically efficient. This is reflected in the wood prop's curve being placed below that of the metal prop and the top curve representing a composite propeller, which is far more efficient than a metal or wooden prop. In comparison, marine propellers have an efficiency of around 56%.

Diagram 10, represents the efficiency for a fixed pitch prop and is similar to the graphs in Diagrams 6 and 7. The sharp angle of the curve is due to the decreasing efficiency at the lower speed ratios (V/nd). Consider the ratio V/nd, if 'V' (true air speed) is zero then the prop's efficiency would also be zero and the aircraft would not move from a stationary position. However, due to the propeller's rotation it still moves a large mass of air rearwards at a low velocity producing 'static thrust'. This is what moves the aircraft from a stationary position. Static prop thrust will be covered in greater detail in the section on 'Propwash-

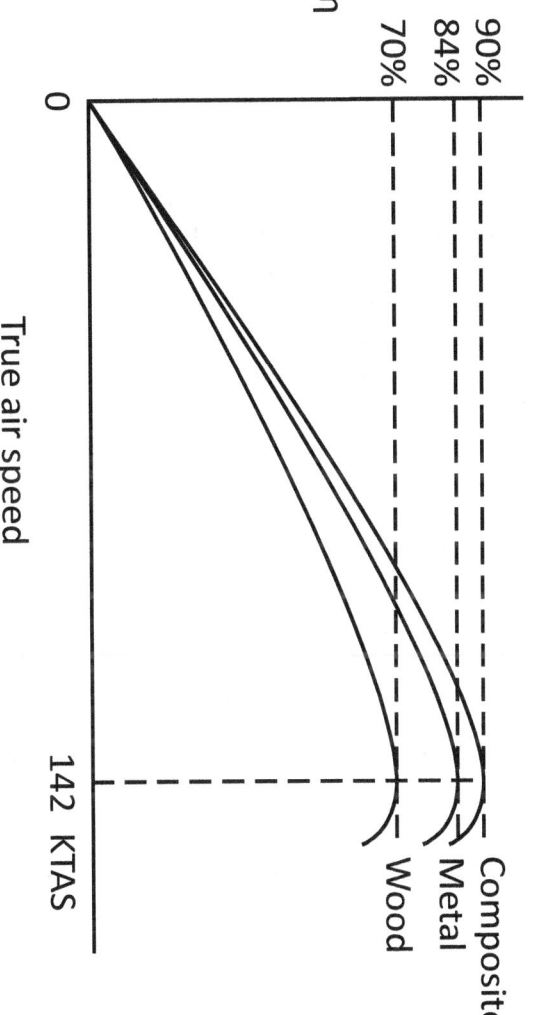

Diagram 10, Efficiency V. True Air Speed

thrust'; that is where the axial momentum theory takes over from the blade element theory.

An alternative formula to find the efficiency of the prop is as follows:

$$\text{Efficiency} = \frac{\text{thrust} \times \text{TAS}}{\text{drag} \times \text{RPM}}$$

This formula is reduced to read thrust/drag ratio (T/D) and TAS/RPM ratio, otherwise known as the speed ratio. Any change in the prop's helix angle due to a change in either RPM or true air speed will increase one ratio and decrease the other by a like amount. The thrust is required to be as high as possible, because greater thrust equates to greater forward speed for a given horsepower. Conversely, drag is required to be as low as possible – less drag gives a higher forward speed for a given thrust; that is, a high thrust/drag ratio is required. Too high a prop tip speed due to high RPM introduces many problems, which will be dealt with later.

Propulsive Efficiency

Propulsive efficiency, not to be confused with propeller efficiency, is the energy imparted to the aircraft, as a percentage of the energy produced by the propeller, or jet engine.

An inspection of Diagram 11, Propulsive Efficiency V. KTAS, compares the efficiency of the different types of aircraft propulsive systems, referred to here as the propulsor, piston-prop engine, turboprop, Propfan and turbofan engines. At the average jet cruise speeds of around 500 KTAS, the efficiency of the turbofan is reaching its peak efficiency while the turbojet's efficiency is still increasing and is good for speeds up to about 2000 knots

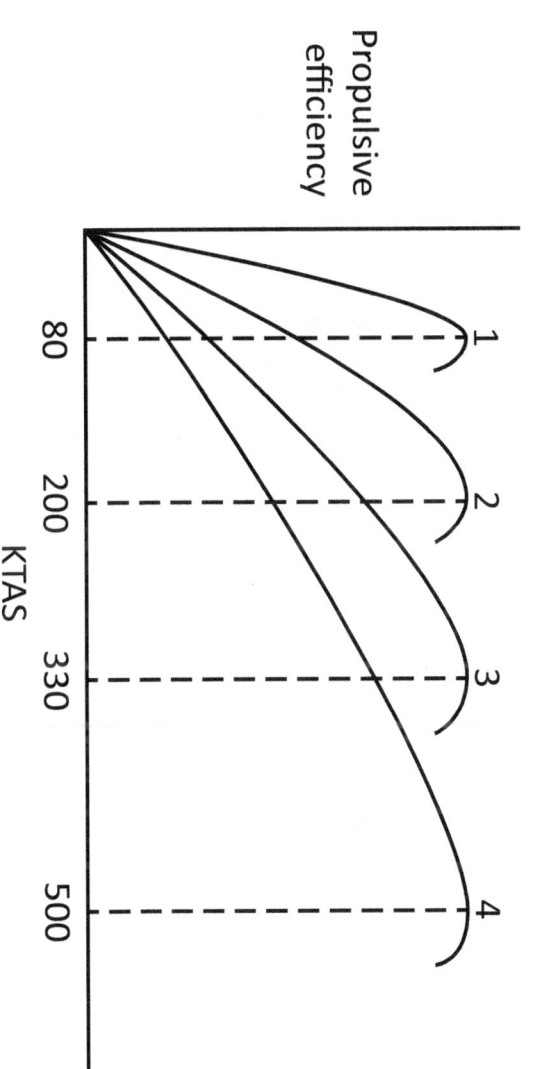

Diagram 11, Propulsive Efficiency v. KTAS

3 – THRUST & EFFICIENCY

and altitudes of 90,000 feet, making it more suitable for military aircraft such as supersonic fighters. The turbofan's limit is reached around Mach 1.0 (575–661 knots where it fits the slot nicely between the turboprop's and turbojet's speed range making it ideal for Bizjets and air transport aircraft. Although the jet engine is very efficient for high cruise speeds at high altitudes, its fuel consumption is uneconomical at low speeds and low altitudes.

At the low end of the speed range the piston-engine/propeller reigns supreme. However, like any airfoil, the prop obeys the laws of aerodynamics and its performance is limited by the constraints of decreasing air density with altitude and the effects of high speed. The prop achieves its greatest propulsive efficiency at around 330 KTAS or so, depending on the type of engine driving the propeller. The different engine types are listed below to match the numbers above the curves on Diagram 11, Propulsive Efficiency V. KTAS, thus:

1. Propulsor
2. Piston-engine
3. Turboprop
4. Propfan and turbofan.

The curves are representative of the approximate average cruise speeds for each engine/propeller combination.

Propulsive efficiency is the product of propeller efficiency and the engine's brake thermal efficiency expressed by the following formula:

$$\text{Propulsive efficiency} = \frac{2 \times V_a}{V_j + V_a}$$

Where V_a = aircraft speed in FPS (feet per second)
V_j = propwash speed in FPS

This also will be covered in the section on 'Propwash-thrust'. For propeller powered aircraft, the propwash velocity during cruise is nearly the same as the aircraft's cruise speed. Therefore, the propeller and propulsive efficiency are both identical.

Power to the Prop

The power output of a piston engine is found by coupling the engine to a test-bed dynamometer. The device measures the torque or turning force of the engine crankshaft in pounds-foot (not to be confused with foot-pounds of work). The torque in pounds-foot can be converted mathematically into brake horsepower by the following formula:

$$BHP = \frac{2\pi \times \text{torque} \times \text{RPM}}{33{,}000}$$

The word 'brake' as in brake horsepower is taken from the dynamometer's alternate name of 'Prony brake'. In practice, the BHP is given to indicate the power of a piston engine. Manifold pressure and RPM are selected by the pilot to produce a required percentage of BHP. For turboprops the term used is shaft horsepower (SHP) or equivalent shaft horsepower (ESHP) if the jet exhaust produces some propulsive thrust.

The maximum BHP for the engine can be plotted on a graph as shown in Diagram 12, Thrust & Power Curves. These are the performance curves familiar to all students of aerodynamics. The maximum BHP is considered here to be a constant, but the BHP produced at any given time does vary under the constraints of increased altitude, temperature, power settings and supercharging. Power

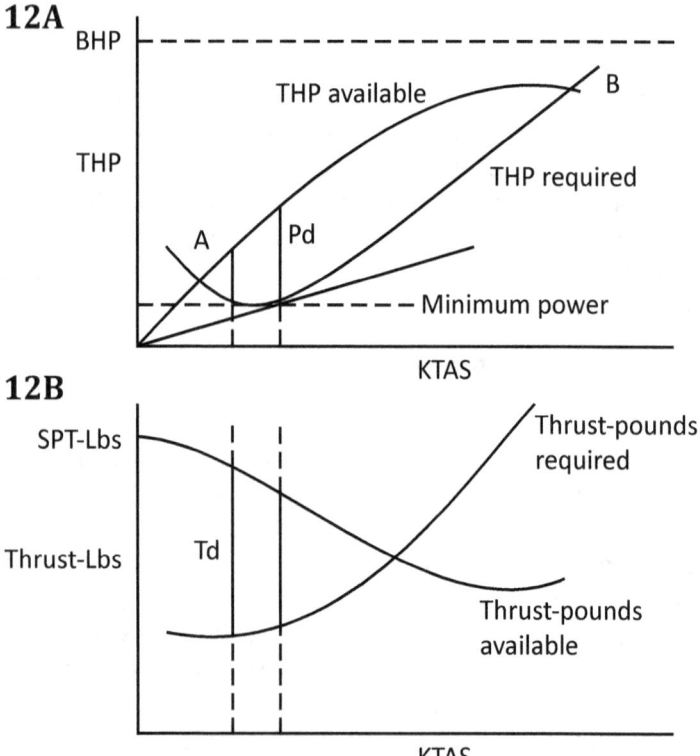

Diagram 12, Thrust & Power Curves

is the rate of doing work; that is, force or thrust times velocity.

The thrust horsepower available curve is plotted on the graph against KTAS after multiplying the BHP by the propeller efficiency. The THP will never be as great as the BHP due to the prop's deficiencies. The performance curves for the THP available, unlike the straight BHP available curve, increases steeply up to the prop's design speed and then reduces again to indicate the propulsive efficiency variation with increasing KTAS. The thrust horsepower required curve is the power required to equal the aircraft's aerodynamic drag. This curve is calculated for various speeds and plotted after being calculated from the following formula:

$$THP = \frac{drag\ (lbs) \times velocity}{factor}$$

$$Or\ THP = \frac{drag \times velocity \times 60}{33,000}$$

Where Thrust = prop thrust in lb
Velocity = FPS, MPH or knots
550 = factor for FPS
375 = factor for MPH
325 = factor for knots

An inspection of Diagram 12A, The Thrust & Power Curves, reveals various aspects of the aircraft's performance parameters. Points 'A' and 'B' on the THP curve shows the minimum and maximum speeds for straight and level flight respectively. The line 'Pd' represents the maximum power differential, or the 'excess thrust horsepower' which produces the greatest rate of climb. A propeller is designed to be most efficient at the aircraft's design cruise speed.

3 – Thrust & Efficiency

Above and below the design cruise speed the prop's thrust and efficiency deteriorate. If the thrust were constant, then the aircraft would achieve its maximum rate of climb at the minimum horsepower required speed, but due to the loss in propeller efficiency, the maximum rate of climb is slightly higher than the minimum power speed.

Moving on to Diagram 12B, shows the curve for 'thrust-pounds available' from the propeller and the second curve represents the 'thrust-pounds required' to equal the aircraft's drag. The prop's thrust is at a maximum at full engine power and zero forward velocity and decreases as the aircraft accelerates. At zero forward speed the thrust is known as 'static prop thrust' measured in pounds (SPT-lbs, on the diagram) when referring to a piston-engine and turboprop aircraft, as opposed to 'static thrust' when referring to jet engines. The prop produces on average between 2–6 pounds of static prop thrust per BHP. On the thrust-pounds curve (Diagram 12B) the line 'Td' represents the maximum thrust differential available from the propeller, as opposed to the maximum excess thrust horsepower shown in Diagram 12A. Notice this speed is slightly lower than that for maximum rate of climb and at this speed the maximum angle of climb will be achieved.

Using the figures given previously for a fictitious aircraft, the thrust horsepower can be found mathematically at the design cruise speed, followed by an alternate method to find the prop's efficiency. Given the engines maximum power of 200 BHP and a prop efficiency of 84% the THP available can be calculated:

$$\begin{aligned}\text{THP available} &= \text{BHP} \times \text{prop efficiency}\\ &= 200 \times 0.84\\ &= 168 \text{ THP available}\end{aligned}$$

Power to the Prop

With the THP and BHP now known, the alternate method to find the prop efficiency is given as:

$$\text{Efficiency} = \frac{\text{THP} \times 100\%}{\text{BHP}} = \frac{168 \times 100\%}{200} = 0.84 \text{ or } 84\%$$

The propeller's efficiency is expressed as a percentage of the ratio of power output to power input. The input is the BHP delivered from the engine to the prop and the output is the thrust horsepower delivered by the propeller. The formulas required for the BHP and THP were given earlier in this section. The thrust force delivered by the prop is found from the next formula:

Thrust = $C_T \rho N^2 D^4$

Where C_T = thrust coefficient
ρ = air density
N = RPM
D = prop diameter

The power required from the engine to turn the propeller is found from the next formula:

Power = $C_P \rho N^3 D^5$

Where C_P = power coefficient
ρ = air density
N = RPM
D = prop diameter

From the above thrust and power formulas a further method can be used to find prop efficiency after cancelling air density (ρ):

$$\text{Efficiency} = \frac{V C_T N^2 D^4}{C_P N^3 D^5}$$
$$= (C_T/C_P) \cdot (V/ND)$$

3 – Thrust & Efficiency

If these formulas are too complex, it can be simplified by a more straight forward formula. It has already been established the ratio of thrust horsepower to brake horsepower equals the prop's efficiency. In addition, thrust power is thrust-force pounds times the aircraft's speed in feet per second, or simply, TV ft-lbs/second. The power required to turn the propeller is BHP times 550 ft-lbs/second. This simplifies the propeller efficiency formula to:

$$\text{Efficiency} = \frac{\text{Power output}}{\text{Power in}} = \frac{TV}{P}$$

Where T = thrust-force in pounds (or kg)
V = aircraft speed in FPS (or m/s)
P = BHP × 550 ft-lb/second (or joules)

Diagram 13, Thrust & Power Coefficients, shows the thrust & power coefficients (C_T & C_P respectively) plus prop efficiency plotted against V/ND. The thrust and power coefficients could be plotted against angle of attack, but being a variable quantity it is more convenient to plot against V/ND. These types of graphs for a family of props are plotted by the propeller manufacturer to determine the values of thrust and power coefficients for any given value of V/ND. Note the curve for Efficiency versus V/ND in Diagram 13, is the same as that included in Diagram 10, Efficiency v. TAS. In conclusion, the propeller's efficiency can be found from various formulas, depending on which factors are available for the calculations.

Power Absorption

The amount of engine BHP a prop can absorb and convert to thrust depends on a variety of factors, including the prop's diameter, number of blades, the blade's aspect ratio,

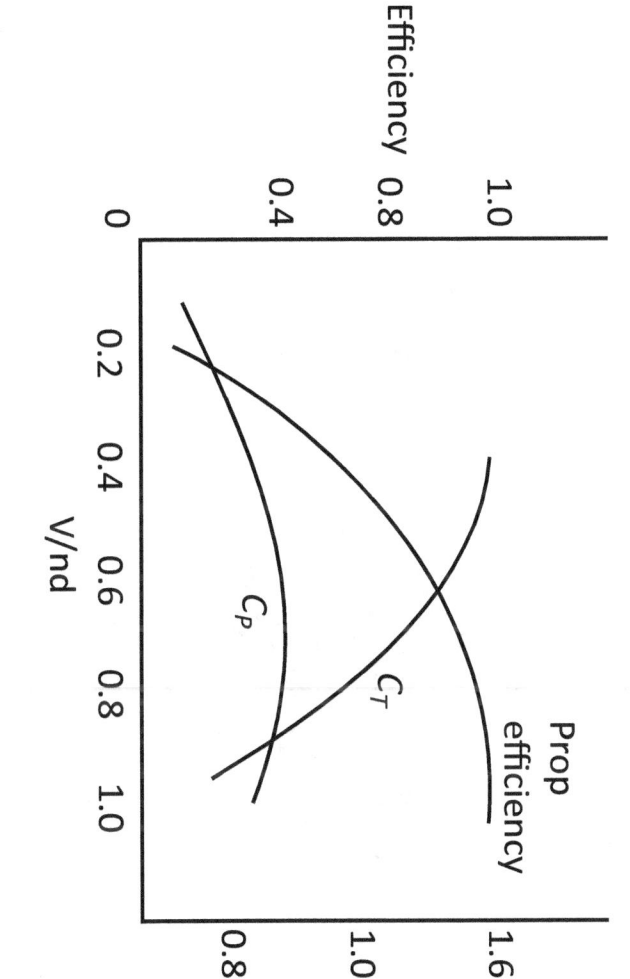

Diagram 13, Thrust & Power Coefficients

3 – Thrust & Efficiency

which all have an affect on the solidity of the prop, while the engine's BHP and the prop disc area determine the amount of prop disc loading.

Activity Factor

Each different type of aircraft can accept a range of engines of varying amounts of horsepower, within certain limits. Likewise, each engine can accept a variety of propellers, also within certain limits. One of these limitations is the propeller's ability to absorb the power provided by the engine. How much power the prop can absorb is measured by the 'activity factor'. In the fictitious aircraft example

The Lockheed P3-C Orion has wide chord, paddle blades with blade root cuffs for greater solidity and improved intake air flow.

presented here, the maximum thrust horsepower available is 84% of the engine's BHP; therefore, the activity factor is 0.84. The activity factor is just another way of expressing the prop's efficiency.

Solidity

Each prop has a maximum limit to the amount of thrust it can produce. If the aircraft designer requires more thrust, then a prop with greater solidity will be required. Solidity is the ratio of total blade area to total prop disc area, which can be found from the following formula:

Solidity = $S/\pi R^2$

Where S = total blade area
πR^2 = prop disc area

A two-blade propeller has a solidity of about 0.08 increasing up to around 0.16 for a four-blade propeller. Solidity, and therefore thrust, can be increased by using a prop with either:

- A greater diameter
- A greater number of blades, or
- Wider chord blades.

The propeller diameter has greater influence on power absorption than the number of blades or blade chord.

Prop Diameter

Reference to Diagram 14, Prop Diameter V. BHP, it can be seen what increase in diameter is required on a two-blade prop if the brake horsepower in increased. If the increase

3 – Thrust & Efficiency

in diameter is unacceptable for any reason, then the next choice is to use a prop with three blades. The fictitious aircraft example with a 200 BHP engine, a two-blade prop of 74 inches could be used. If the 200 BHP engine was replaced with a 250 BHP engine, it would require a prop of 77 inches. But, if this diameter is too large the next choice is a three-blade prop of 74.5 inches. [Note this diagram is not a true one but hand drawn to illustrate the point being made].

The prop diameter on a single-engine aircraft may be limited by ground clearance or, on a multi-engine aircraft, clearance between the prop tips and the aircraft's structure and the ground, which must be taken into consideration. Propeller tip clearance must comply with the certification regulations; for example, the American FAA regulations require a minimum clearance between the tips and the ground of seven inches (17.78 cm) for nose wheel aircraft and nine inches (22.86 cm) for tail-wheel aircraft.

Number of Blades

The solidity of the prop can be increased to absorb more power by increasing the number of blades mounted on the prop. However, there is a limit to the number of blades that can be used. Five-blades are the accepted maximum number for a metal prop and now, up to eight for a composite prop; however, that is encroaching into the realm of Propfans. Six or more blades can be used on a composite prop due to their lower weight and superior efficiency over metal blades. The reason for the restriction on the number of blades is the air cascading over the following blades causing interference drag and reducing efficiency. A greater number of blades not only produce more thrust, but also more drag at the idle power settings

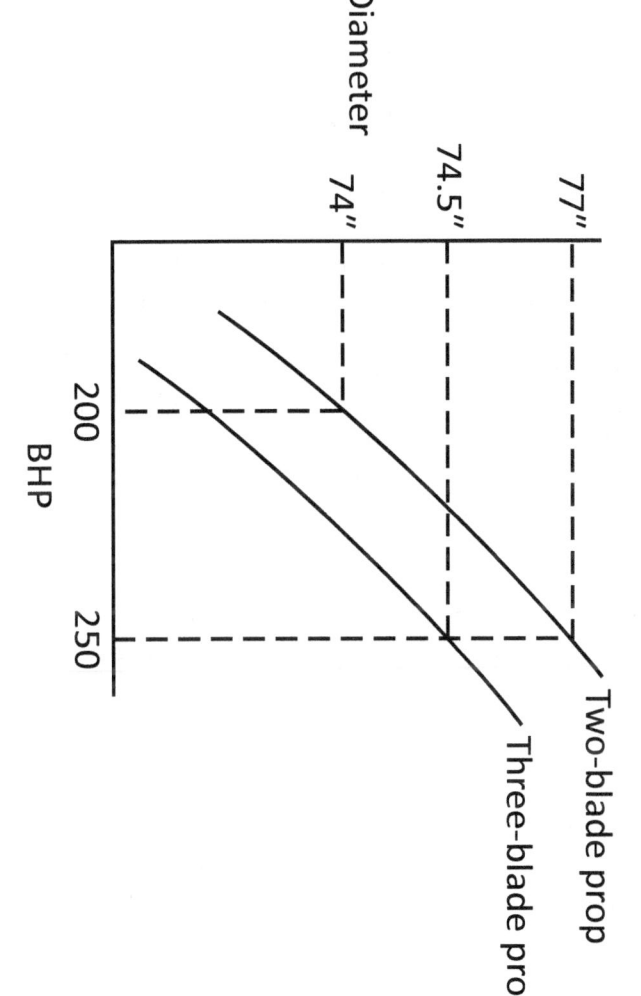

Diagram 14, Prop Diameter v. BHP

3 – Thrust & Efficiency

which can be used to advantage by acting like an air brake. The cruise speed can be quickly reduced to approach speed combined with an increased rate of descent when desired.

Using three or more blades on a prop in place of a two-blade prop can be aesthetically pleasing, and hint at higher performance although this may not always be so. For example, the Cessna 207 light aircraft maybe equipped with either an 82 inch two-blade prop or the optional three-blade prop with a slightly smaller diameter. However, to quote from the Cessna 207 owner's manual (POH), "There is no significant performance change with the three-blade propeller". In contrast to this, when Piper Aircraft developed the Piper Malibu, they found a two-blade prop gave the aircraft better performance in the cruise than that obtained from a three-blade prop. This may be attributed to prop blade loading; during low-speed operations such as take-off and climb, the three-blade prop has a lower prop loading. It can absorb more power and increase the static prop thrust, which in turn, will increase the initial acceleration during the take-off roll and improve the rate of climb. In effect, the three-blade prop does not have to work so hard. It is more efficient than a two-blade prop at lower speeds. At the higher speeds of cruise flight, the thrust available and hence the prop blade loading is lower still. This makes it more difficult for a three-blade prop to maintain a sufficiently high lift/drag ratio along with high efficiency. This may become evident with a slight loss in cruise speed, especially at higher altitudes. The aerodynamic drag can be expected to be less when the prop blade loading is low, which is usually found on three-blade props at cruise speed. On the other hand, a lightly loaded three-blade prop can produce as much drag as a two-blade prop. A two-blade prop with less solidity will suffer from too high prop blade loading and will be less

NUMBER OF BLADES

The Lockheed C-130J with six-blade scimitar props in the distance, is overshadowed by the older Lockheed C-130H with four-blade, square-tip propellers.

efficient during take-off and climb. During cruise flight, the prop blade loading will be relatively higher than a three-blade propeller. This will enable the two-blade prop to produce a better lift/drag ratio and greater efficiency during cruise flight.

The number of blades also affects the vibration produced by the prop. It is a well-known fact amongst pilots that a prop with three or more blades runs smoother than a two-blade propeller. It was mentioned above; increasing the number of blades reduces prop blade loading and therefore less thrust per blade. This results in the vibration's frequency being raised resulting in a reduction in the amplitude of vibration. When the prop produces higher vibration frequencies, the amount of vibration transmitted through the airframe to the occupants is less discernible and is felt as a smoother running engine.

A close up view of the six-blade scimitar propellers of the Lockheed C-130J Hercules.

There are many variables involved in determining a propeller's efficiency. The aircraft designer must decide at what true air speed the two-blade prop becomes a better option than a three-blade prop. It was mentioned above; the three-blade prop produces better performance and efficiency during take-off and climb. Therefore, if the performance can be maintained right up through the normal cruise speed, then the three-blade prop is the right choice for that aircraft type. If the three-blade prop loses efficiency before the aircraft reaches its design cruise speed, then the designer is more likely to opt for a two-blade propeller. It all depends at what true air speed the three-blade prop becomes more efficient than a two-blade propeller mounted on that particular type of aircraft. Of course, the same argument applies when we compare a three-blade propeller with a four-blade unit, or a four-blade prop with a five-blade prop, etc.

In conclusion, a three-blade propeller is better for take-off and climb performance with the added benefit of reduced noise and less vibration, but a two-blade prop may, or may not, produce better cruise performance. As with most things aeronautical, it is again a matter of compromise.

Aspect Ratio

The blade's chord along with its length determines the aspect ratio. The aspect ratio is defined as the ratio of prop radius to prop chord (R/C). A high aspect ratio prop blade – one with a narrow chord – is generally more efficient than a low aspect ratio or wide chord blade. A high aspect ratio blade shares the same characteristics with a high aspect ratio wing; the strength of the trailing edge and tip vortices is reduced. Therefore, the induced drag is also

The Scottish Aviation Twin Pioneer of 1955 vintage has three, high-aspect ratio prop blades, de-iced with a curved leading edge and a trailing edge cuff. It is housed in the RAF Cosford Museum, England.

reduced, which increases the prop's lift/drag ratio and hence, efficiency. But there is a limit to the blade's high aspect ratio; problems may occur with blade stall and the strength of the blades due to the various forces acting on them. [See Prop Stress].

A prop with a low aspect ratio blade is known as a 'paddle blade'. It has greater solidity and a higher activity factor, and will absorb more engine power than a high aspect ratio blade. But its efficiency may be reduced due to the blade wake affecting the thrust produced by the flowing blades. A wide chord blade also places more stress on the pitch change mechanism.

Prop Blade Loading

Prop blade loading is equal to the BHP divided by the prop blade area (BHP/PBA), as opposed to prop disc loading, which is BHP divided by prop disc area (BHP/PDA).

$$\text{Prop blade loading} = \frac{\text{BHP}}{\text{Prop blade area}} = \text{HP/sq.ft}$$

Increasing the number of blades can reduce the prop blade loading for a given prop. It is akin to wing loading where the reduced strength of air circulation and a reduction in the air flow velocity over the prop blades causes a reduction in prop blade loading (or wing loading). The reduced air flow over the prop blades may at first be puzzling when examining Diagram 16, Forces in Cruise Flight, which shows the 'relative air flow' to be dependant on the RPM and forward velocity. As the air mass flow approaches the prop disc, the velocity in front of the prop disc increases and then decreases behind the prop disc; with increased blade area (reduced blade loading) the air circulation over the blades is reduced.

3 – Thrust & Efficiency

Prop Disc Loading

Prop disc loading is defined as the engine's 'BHP divided by the prop disc area'. If the propeller's diameter is increased, it will lead to an increase in the prop disc area, which will reduce the prop disc loading and in turn, increase the propeller efficiency. The prop disc loading can be reduced by either increasing the prop diameter or by reducing the engine's BHP. If the BHP is increased and a prop of the same diameter is used, it follows the prop disc loading will be increased. However, if the increase in prop disc loading is too great, a loss in efficiency will result. This is due to the increased air pressure at the rear of the prop disc leaking around the propeller tips causing an increase in prop tip vortices and induced drag. The same affect occurs on a wing that is too short for the aircraft. In fact, the prop disc loading has the same affect as the aircraft's wing loading.

The value of the prop disc loading can be found given the engine's BHP and the prop's diameter, or better still, its prop disc area, which for a 74 inch diameter prop is found to be 29.86 square feet. The prop disc loading is then found from the formula:

$$\text{Prop disc loading} = \frac{\text{Brake horsepower}}{\text{Prop disc area}}$$

$$= \frac{200 \text{ BHP}}{29.86 \text{ sq.ft}} = 6.7 \text{ HP/sq.ft}$$

The power absorbed by a fixed-pitch propeller will vary as the cube of the RPM change (RPM^3) depending on the air density and RPM. A given engine power is produced by any one given RPM, air density being constant. The maximum RPM of a modern light aircraft piston-engine is usually limited to around 2700 RPM, mainly due to the noise and compressibility caused by the high propeller tip

Prop Disc Loading

speed. Some relatively recent production models run even slower usually around 2500 RPM maximum. Most engines could run up to about 3600 RPM before destruction occurs, but it is the prop tip speed that determines the maximum allowable engine RPM, indicated by the red line on the engine's tachometer. A propeller reduction gear with a fixed gear ratio will then be incorporated in the drive between the engine and propeller. As far as the engine is concerned, running at higher RPM is advantageous because the greater number of power strokes per minutes produces greater power. The Lycoming TIO-540 engine is a good example here; this is a direct drive engine producing 380 BHP at a red line of 2900 RPM. The geared version of this engine, the TIGO-541, produces its maximum power of 425 BHP at 3200 RPM. Both engines are identical except for the reduction gear on the TIGO-541 engine. The propeller on the geared engine can absorb the greater horsepower and therefore, produce greater thrust while maintaining the prop tip speed within acceptable limits.

With the prop of a geared engine turning at a lower RPM than a direct drive prop, the blades will meet the airflow into the prop disc at a much lower speed. They would therefore do less work in producing thrust, resulting in a loss in efficiency. To overcome this loss, the prop's solidity is increased by using more blades or wider chord blades, or by increasing the prop's diameter. However, increasing the diameter too much results in too a high tip speed, which was the reason for using reduction gear in the first instance. This emphasises the need to match the prop to the engine's BHP and the aircraft design air speed. On a piston engine, the reduction gear will have a ratio of around 3:2, but for turboprop engines due to their inherent design, have an operating speed of 10,000–15,000 RPM (or even higher depending on the engine design) require a much greater reduction gear ratio. The Rolls Royce Dart engine for example, has ratio of 10.75:1. It must now be emphasised here, it is the engine that is geared, not the propeller. A geared piston engine can be recognised by its designation. For example, the letter 'G' in Lycoming's TIGO-541 engine indicates the engine is geared.

It has been determined thus far that the maximum propeller efficiency occurs when the prop produces the maximum thrust/torque ratio. Also considered were the factors, which determine how much engine power, or torque, the prop absorbs and transmits as thrust energy to the propwash.

Propwash Thrust

Throughout this book we have followed along the lines of the 'blade element theory'. At this point we diverge briefly to consider the 'Rankine-Froude axial momentum theory'; this theory gives a clearer explanation of the energy change in the propeller's propwash.

The marine engineer, R. E. Froude (1846–1924), introduced the idea of the prop disc, which became known as Froude's Actuator Disc where the air mass on passing through the prop disc experiences a sudden rise in pressure without affecting the increasing the propwash velocity. The axial momentum theory assumes the prop thrust to be evenly distributed through the propeller disc, which just isn't true. In fact the thrust varies from zero at the hub to a maximum at the 0.75 radius station and reduces again to zero at the prop tips. It must also be realised no propeller achieves the efficiency an actuator disc implies. This fact will be disregarded for the present time and it will be assumed the thrust to be evenly distributed across the prop disc, while discussing the axial momentum theory.

Propwash Thrust

As the propeller rotates under normal operating conditions, it sucks air from in front of the prop disc causing a low pressure area. The air mass, known as the 'inflow velocity' (V1) accelerates through the prop disc into an area of increasing velocity behind the prop and experiences a rapid rise in pressure as it does so. It is the difference in pressure between the front and rear of the prop disc, caused by the change in the air mass momentum that produces thrust. The air mass now called the 'outflow velocity' (Vo) continues to accelerate reaching its maximum velocity some distance behind the propeller. The final maximum velocity is equal to the aircraft's rue air speed plus double the propwash velocity (aircraft speed + 2V). It should now be obvious from the above, half of the propwash velocity increase occurs in front of the prop while the other half of the speed increase occurs behind the propeller. As the speed of the propwash increases to its maximum value, the propwash also contracts to a smaller diameter than the prop disc itself, in compliance with Bernoulli's Theorem. The point of constriction is known as the 'Vena Contracta'. The air mass flowing through the prop disc should be considered as a three-dimensional stream tube.

The propwash velocity/aircraft velocity ratio (v/V) is known as the 'inflow factor' (a), which increases with an increase of propwash velocity. Propeller efficiency will be greatest when the propwash velocity is close to the aircraft's true air speed (V), that is, the greatest efficiency is achieved at a small value of v/V. When the propwash velocity equals the aircraft velocity, the value of the inflow factor (a) is equal to one. At other speeds, where the propwash velocity is less than the aircraft's velocity (which is the usual case) the value of the inflow factor will be less than one, which ties in with the 'ideal efficiency'

3 – Thrust & Efficiency

of the prop, where the thrust is related to aircraft's speed and the inflow factor. The ideal efficiency, or Froude's Efficiency, can be calculated from the formula known as the Froude's Equation:

$$\text{Ideal efficiency} = \frac{T \times V}{T \times (V + v)}$$

Where T = thrust factor
V = aircraft velocity
v = propwash velocity

The propwash velocity (curve A) is plotted against the aircraft's true air speed (curve B) in Diagram 15, Propwash Velocity v. KTAS. This shows the static prop thrust to be 90 knots at full throttle while the aircraft is stationary. As the aircraft accelerates, the propwash will also accelerate but not as quickly, until a point is reached where the aircraft's velocity and propwash velocity coincide. This occurs at the maximum level flight speed of the aircraft. Stated simply, the thrust equals the mass of the air multiplied by the propwash velocity minus the aircraft velocity.

It was stated earlier, "It is the difference in air pressure between the front and rear of the prop disc that produces thrust". However, this is only part of the story. As the aircraft travels forward, the air mass well ahead of the prop disc is assumed to be stationary; on approaching the prop disc and passing through it, the air mass accelerates and gains momentum. Thrust is a result of this change in momentum and the greater the change, the greater the thrust. Momentum is the product of mass multiplied by velocity (MV). A large mass of air flowing at a small velocity can produce the same amount of thrust as a small mass of air flowing at a high velocity. The former is the better of the two options because it requires less work to produce

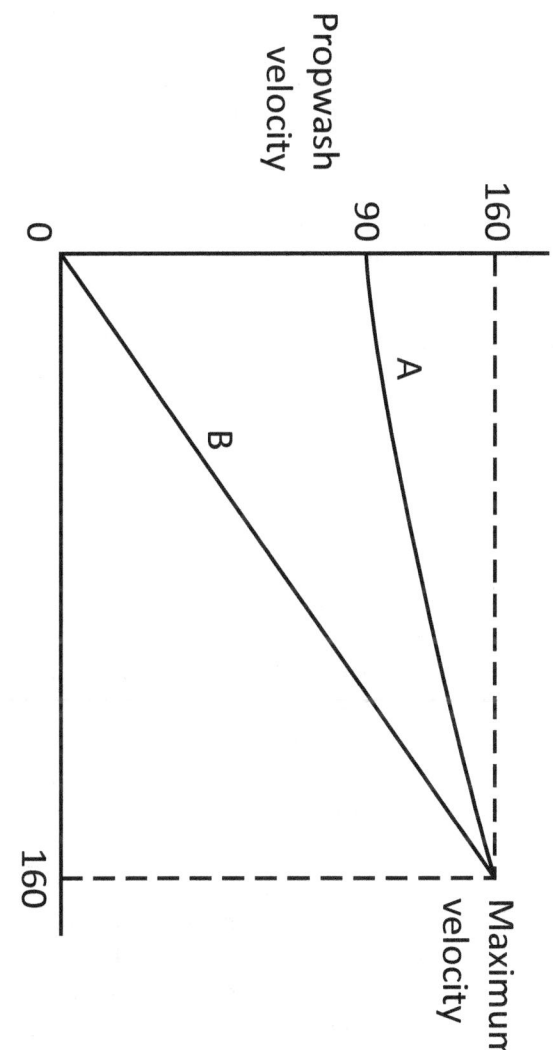

Diagram 15, Propwash Velocity v. KTAS

3 – Thrust & Efficiency

the same amount of prop thrust. The following can show this. The propwash gains kinetic energy ($\frac{1}{2}MV^2$) due to the increase in momentum (MV) as it passes through the prop disc. The kinetic energy represents the work done by the prop in accelerating the air mass from zero velocity. The kinetic energy of 'M' slugs moving at a velocity of V feet/second is equal to $\frac{1}{2}MV^2$ ft. lb. It can be seen the same thrust and momentum can be achieved from either one slug given 10 ft/second acceleration or, ten slugs an acceleration of one feet/second. Using the above formula ($\frac{1}{2}MV^2$) it is shown:

1. One slug given 10 ft/sec = ½ × 1 × 10^2 = 50 ft.lbs
2. Ten slugs given 1 ft/sec = ½ × 10 × 1^2 = 5 ft.lbs.

Alternately, the metric formula can be used using kilojoules where:

M = mass, kilograms (kg)
V = velocity, metres/second (m/s).

Therefore, the second line (2) is more efficient because it wastes less energy and produces the same momentum by accelerating a large mass of air at a low velocity. From this result it is proven a large propeller is more efficient at the relatively lower aircraft speeds than a jet engine, which accelerates a relatively small volume of air at a much higher velocity. The reverse becomes true at higher aircraft speeds where the jet engine has superior efficiency, due to other factors not related to propellers. [See Diagram 11, Propulsive Efficiency v. True Air Speed].

The static prop thrust produced when the aircraft is stationary can be calculated from the following formula:

Static prop thrust = PDA × V1 × ρ × Vo

The cruise thrust can be calculated from a similar formula:

Cruise thrust = PDA × (V + V1) × ρ × Vo

Where PDA = prop disc area, sq.ft (or sq. m)
V = aircraft velocity, FPS (or M/s)
V1 = inflow velocity, FPS (or M/s)
Vo = outflow velocity, FPS (or M/s)
ρ = air density, slugs/cu.ft (or kg/m^3)

Note, when using Imperial units the answer is in pounds of thrust and for SI units the answer is in Kilogram force. The difference in the formulas between the static prop thrust and cruise thrust is the addition of the aircraft velocity (V) in the cruise formula. When referring to the propwash velocity, unless otherwise stated the velocity in the axial direction only is considered. The helical velocity (or race rotation) as it is known, is ignored.

The location of the propeller on the aircraft is an important consideration for the aircraft designer. If it is placed too close to the airframe or engine nacelle, thrust and efficiency can be reduced and also, the propwash velocity will reduce and its pressure will rise. This fact becomes relevant when debating the advantages and disadvantages between pusher and tractor propellers: some people claim pushers have greater efficiency. But do they? The air mass flowing through the prop disc is accelerating causing a pressure differential, by lowering the pressure in front of the prop and raising it behind, as mentioned previously. The propwash from a single-engine tractor prop flows over the entire fuselage increasing the parasite drag in proportion to the greater pressure gradient created by the propeller. This can amount to a 5% increase of fuselage drag equating to prop efficiency about 4% lower than a pusher prop, thus favouring a pusher as being slightly more efficient. If the prop is now placed at the far end of the fuselage as a pusher, the 5% increase in parasite drag mentioned above, is now cancelled. However, it is not all good news! The decrease of air pressure in front of the pusher prop's disc has the effect of lowering the pressure gradient on the aft portion of the fuselage resulting in an increase of drag in that area. This is equivalent to a loss in prop thrust of about 5% and a loss in propeller efficiency of 2–3%. The net result is, tractor and pusher props come out about even as far as efficiency is concerned.

What has been said above regards fuselage mounted tractor and pusher props, also applies to wing mounted engines on multi-engine aircraft. The engine cowling causes a disturbance to the passing propwash, which reacts back on the propeller as interference, which reducing the prop's apparent pitch and efficiency. Pusher props are affected by the reduced propwash velocity and interference caused by the engine cowling in front of the prop. The reduced propwash mainly affects the inner portion of the propeller blades, leaving the outer portion unaffected as it operates at a higher rate of advance than the inner portions of the blades.

With the exception of the above paragraph, it has been assumed the propeller to be 'free standing', and unaffected by the presence of the fuselage or engine nacelle behind the prop: the thrust is then referred to as 'free thrust'. However, if we take into consideration the presence of the fuselage or nacelle and its affect on the propeller propwash, we then refer to the disturbed propwash as 'apparent' or 'gross thrust'. Going one step further, if the drag is subtracted from the gross thrust (caused by the propwash flowing over the fuselage or nacelle) the term is 'propulsive' or 'net thrust'. Propulsive thrust is always a constant fraction less than the apparent thrust due to the drag being proportional to the 'propwash velocity squared'. That is the theory according to R. E. Froude... we now resume with the blade element theory.

One advantage a propeller has over a jet engine is the addition of the propwash flowing over the parts of the wing and empennage (tailplane). The total lift produced by the wings is influenced by the total slipstream over the whole aircraft plus the wing lift enclosed within the propwash. The total amount of lift can be varied within limits, by variation in engine power settings and thus changes in propwash thrust with the aircraft at constant speed. [The jet aircraft must increase air speed to increase lift, and due to the aircraft's inertia, this takes time]. Variations in propwash and wing lift can be used to advantage during the approach to land when the aircraft may experience a rapid sink. An increase in engine power will increase the propwash-flow over the wing lift and thus, increase lift to stop the sink. With power on, another advantage is the additional lift that lowers the stalling speed by 5–10 knots, depending on the aircraft type.

Prop Blade Drag

The propeller acts like any airfoil moving through the air, it produces an aerodynamic reaction due to its shape, angle of attack and velocity. The total reaction can be divided into vector components of lift and drag. It is the drag component of interest here.

Examination of Diagram 16, Forces in Cruise Flight, shows when the prop is operating at its maximum lift/drag ratio it is producing the most lift for the least drag. It follows, the most thrust for the least amount of engine power used (the maximum thrust/torque ratio) will also be achieved at the maximum lift/drag ratio condition.

3 – Thrust & Efficiency

Therefore, the prop will be operating at its maximum efficiency.

With the prop advancing through the air on its helical flight path, there is an upwash of air in front of the blade and a downwash behind, the same effect that occurs on an aircraft wing. The net result of this upwash and downwash is a general downwash of the relative air flow over the blade. Because the total reaction is at right angles to the relative air flow, it will be tilted rearwards from the vertical relative to the blade element. The horizontal component of the total reaction represents the propeller's induced drag, or to use the modern terminology, trailing vortex drag. [See Diagram 1, Airfoil Terminology].

The pressure differential between the propeller blade's top and bottom surfaces causes the air flowing over and under the blade to meet at the trailing edge at an angle to

The Bristol M1c Bullet has a giant-size prop spinner looking very much like a doorbell. This aircraft is located in the RAF Hendon Museum, London.

each other known as the 'rake angle'. This causes a vortex sheet to emanate from each prop blade. High aspect ratio blades produce a smaller rake angle and therefore less induced drag than low aspect ratio blades. This again, reflects the superior efficiency of high aspect ratio blades. The flow around the blade from high to low pressure (rear to the front surfaces) causes the tip vortex to be stronger and cause more drag than the trailing vortex sheet. A similar vortex emanates from the blade root and this rotates in the same direction as the prop, while the tip vortex rotates in the opposite direction. The helical vortex sheet flowing off each blade affects the following blade by causing a disturbance in the air flow pattern resulting in a loss in efficiency: the greater the number of blades, the greater the disturbance. Using more blades negates the advantage of greater solidity, due to the increased flow disturbance.

On an aircraft wing, the vortex drag can be reduced by using an elliptical wing planform or by using wing taper, which has the same effect. However, the planform of a propeller blade has to be designed by calculation and is not necessarily elliptical. This is due to the difference in speed between the blade's root and tip affecting a different amount of air mass in a given time. This problem is partly alleviated by using scimitar shaped blades as found on new generation turboprop aircraft. Scimitar blades are designed with extra chord width around the 50% prop radius station. At low speed during take-off, more of the thrust is produced in this area of the blade. As the aircraft's speed increases the major part of the thrust is produced further outboard along the curved span of the blade providing better efficiency at higher speeds where the effects of compressibility are delayed reducing drag and noise. This effect is akin to a swing-wing fighter aircraft.

The pointed spinner of the Curtiss P-40 Kittyhawk contrasts with the blunt spinner on the Albatross DV.a.

Prop Blade Drag

The propeller drag is caused not only by the blades, but also to a lesser degree, by the prop boss or shank. The blade roots, boss or shank cause profile drag due their inherent thickness. This is one reason for installing a prop spinner, to smooth the air flow over the drag producing area of the prop. At low speed ratios, the loss in efficiency caused by drag is around 10% rising to about 29% at higher speed ratios. The power required to overcome the profile drag is known as the 'profile drag power loss'. Because aerodynamic forces are proportional to the square of the speed, it would appear obvious the thrust and torque would be at a maximum at the propeller tips where the rotational velocity is the greatest. However, this is not so. Due to tip losses caused by the spanwise flow along the blade towards the tips and also the effects of compressibility, thrust and torque values reach a maximum around the 75% prop radius station and decrease towards the tip. The blade chord is at a maximum around the 75% station for this reason, and this is the location of the minimum drag coefficient, as opposed to the maximum drag coefficient, which occurs at the blade tip due to induced drag and also at the blade root due to form drag, as mentioned above. The overall drag coefficient will remain approximately constant if the prop tips are not affected by compressibility.

Sir Isaac Newton's Third Law of motion states, "For every action there is an equal and opposite reaction". In providing power to turn the propeller, the engine produces a torque component, which is a force acting in the same plane and direction as the propeller rotation. At a constant power setting, the engine torque is balanced by the equal and opposite force of propeller torque, in accordance with Newton's third law. On an aircraft with a fixed-pitch propeller, the engine RPM will remain constant as long as these two forces remain in balance. A change of power setting or aircraft speed will change the value of the engine torque or propeller torque respectively. This will cause a change in the engine RPM. The same applies to a constant-speed prop but the CSU works to maintain a constant RPM masking the changes in prop or engine torque, which varies as the RPM squared. The total drag forces or torque of the propeller act through the centre of pressure of each prop blade. The prop torque can be found from the following formula:

Prop torque = $k_Q \rho N^2 D^5$

Where k_Q = torque
ρ = air density
N = RPM
D = Diameter

Propeller Icing

Propeller icing will form on the airframe or propeller when flying in cloud or rain with ambient temperatures below 0 degrees Celsius (32 °F) down to temperatures around minus 40 degrees Celsius. Between 0 °C and minus 20 °C, icing will be most severe with glaze type of icing. From −20 °C to −40 °C rime ice will be more prevalent and below −40 °C icing is less likely to occur, but is still a possibility. Icing will form on those parts of the aircraft with relatively sharp or protruding items such as the wing's leading edge, aerials, struts and of course, the leading edge of the propeller blades. The weight of the accumulated ice is less of a hazard than the adversely modified airflow over the prop blade. It only takes a small amount of ice to modify the shape of the blade's leading edge and degrade performance thus causing a reduction in prop thrust and efficiency. The associated drag will reduce the rate of

Propeller Icing

A USAF Convair T-29B Flying Classroom with square prop tips and de-icing boots on the props leading edges. This aircraft is stored at the Pima Air & Space Museum, Tucson, Ar.

climb and cruise speed. In addition, if the prop has icing problems, then the wings are sure to be iced up as well. In theory, the aircraft structure should ice up before the prop blades. This is due to the blade tip's high speed causing kinetic heating, which increases the temperature of the blades: the heat rise being approximately proportional to the square of the speed (prop RPM). The blade's inner portions will be rotating at a slower speed than the tips and therefore, will experience less of a temperature rise due to less kinetic heating. This accounts for the electric de-icing heater mats being positioned on the inner leading edge of the blade only and not extending to the tips.

Uneven accumulation or shedding of the ice will put the blades out of balance causing severe vibrations, as opposed to a rough running engine. An 'ice plate' may be mounted on the side of the fuselage on twin-engine aircraft in the prop's plane of rotation, for reinforcement of the fuselage

3 – Thrust & Efficiency

skin against shedding ice strikes. In 1934, B.F. Goodrich pioneered the system of pulsating rubber de-ice boots on the wing's leading edge. Ice protection for the prop blades followed later in the form of electric heater mats, deice boots and a chemical system. If the blade's leading edge is rebated to take the heating element, it is then known as a 'rebated blade'. Propellers with the electro-thermal system installed are commonly referred to as 'hot props'. The chemical system uses Ethylene Glycol, or similar fluid, which is also used in the cooling systems of liquid-cooled engines. The de-ice fluid is dispersed via a slinger ring mounted around the prop hub inside the spinner and centrifugal force carries the fluid along the blade via the ridges in the rubber boots.

As far as the electrical and chemical systems are concerned, there is no difference between anti-ice and de-ice systems. It is all a matter of timing, anti-ice prevents

The Short Belfast T1 turboprop transport with de-iced props. This Belfast is on show in the RAF Cosford Museum, England.

Propeller Icing

and de-ice cures. If icing is expected, prevention is better than a cure, so turn on the anti-ice system early before the ice has a chance to buildup to a dangerous level. If you have the misfortune to experience prop icing with no anti-ice system on board, it maybe possible to remove the prop ice by flexing the blades using centrifugal force. This can be achieved by reducing the engine speed to around 2200 RPM with the propeller pitch control, then quickly move the prop control to fine/flat pitch. Several cycles maybe required to restore the prop to smooth running. After clearing the ice, or if icing is expected, run the engine at a higher RPM than normal to reduce the chance of ice forming. Finally, one last word on prop icing: keep the prop blades smooth and clean and apply a coating of silicone spray, it just makes it that little bit harder for the ice to cling to the blades.

4 – Effect on the Aircraft's Stability

The previous section covered propeller drag and introduced prop torque. Propeller torque is now covered in greater detail, along with its associated propwash force, precession, asymmetric disc loading, 'P' factor and the effect these have on the aircraft's stability. These factors are mostly de-stabilising, however in some cases they can enhance the stability of the aircraft.

Prop Torque Force

The engine torque will produce an equal and opposite torque reaction at the propeller creating a turning moment, which will tend to rotate the aircraft around its longitudinal axis in the opposite direction to the prop's rotation. With a 'right-handed' prop, this will cause the aircraft to rotate or roll to the left, in accordance with Newton's Third Law of equal and opposite reaction. This can present as a problem during take-off due to asymmetric loading on the undercarriage, where the left-hand wheel is pressed down on the runway more so than the right-hand wheel. This excess pressure results in wheel drag and in turn, causes the plane to yaw to the left. For most modern aircraft types, the effect of torque and the accompanying roll and yaw in flight can be considered negligible and is easily corrected by use of the controls. Pilots of tail-wheel aircraft, especially World War II fighters with their greater power/weight ratios, have considerably higher torque forces to contend with. A pilot who is not 'ahead' of his/her aircraft with a high power/weight ratio, could experience a torque roll on take-off, or during a go-around that could end with catastrophic results. The earlier versions of the Supermarine Spitfire were equipped with the Rolls Royce

4 – Effect on the Aircraft's Stability

Merlin engine that rotated the prop clockwise, while the later versions of the Spitfire from the Mark 12 onwards were equipped with Rolls Royce Griffon engines, which rotated the prop anti-clockwise. Pilots converting from the earlier 'mark' of Spitfire to the later models had to be ready to counteract opposite torque forces with the rudder pedals. The torque forces are at a maximum during full power operations such as during take-off and climb out, but the force can be considered as zero during a descent with the engine throttled back to idle setting.

Prop Location

The decision on where to place the propeller and engine unit on the aircraft is a complex and important choice for the aircraft designer. For a single-engine aircraft should it be a tractor or pusher prop? There are many arguments for and against either layout. The Wright brothers chose

The Beech 2000 Starship with twin five-blade pusher props, located at the Pima Air & Space Museum, Tucson, Ar.

pusher props on their Wright Flyer 1, and this arrangement was popular with other aircraft designers that followed. One reason early types of pushers were designed as such was to keep the propeller clear of forward firing guns. The prop's location can also affect the aircraft's stability, which will be covered shortly. The centre of gravity range on other types determined the prop's location and again on other types, maybe it was just the designer's choice. Since the WW 1 era, the trend had been more towards the tractor arrangement, with a few exceptions along the way.

The USAF's Convair B36 long-range bomber must be the world's largest pusher aircraft ever built. It is powered by six Pratt & Whitney Wasp Major R-4360 piston-engines of 3500 BHP each, driving 19-ft. pusher props. Later models had the addition of four turbojet engines to cope with an all up weight of approximately 310,000 pounds With a wingspan of 230 feet, (larger than the Boeing 747B's span of 195 feet 8 inches) it was at the time of its introduction the world's largest aircraft with a first flight on 8 August 1946. The futuristic looking Beech Starship 2000 introduced in 1986 as a business aircraft, is one more relatively recent pusher type. The twin Pratt & Whitney PT-6A turboprops powered pusher props are ideally located on the trailing edge of the wing, to help place the centre of gravity well aft. The Italian Piaggio P.180 Avanti is a similar aircraft in layout to the Starship and commercially, a greater success. It was introduced in the same year as the Starship – 1986 – and is still in production at the time of this writing in 2014.

The Lake Buccaneer amphibian is another pusher type of unusual design. The single pylon mounted engine above the fuselage could well have been installed as a tractor unit. Why did the designer choose a pusher arrangement? Was it for aerodynamic reasons, or maybe to keep the prop away from spray during water take-offs and landings? The

4 – Effect on the Aircraft's Stability

The Convair B-36J Peacemaker, the world's largest aircraft with pusher props. It has six P&W piston engines and four jet engines! The B-36 in this photo was the last one built, in 1956. It is on show at the Pima Air & Space Museum, Tucson, Ar.

majority of prop powered aircraft, with the exceptions mentioned above, are now designed and built as tractor prop aircraft, with the pusher design being an exception to the rule. However, I have digressed, so to continue...

The location of the propeller/engine affects the plane's stability due to its position and the presence of the propwash. Although tractor props are more common, rear mounted pusher props enhance the stability. If the aircraft experiences a yaw for any reason or other, say to the left, the propwash will be deflected the opposite way to the right, and will pull the aircraft's nose back to the right aiding directional stability. [Imagine the propwash acting in the same sense as the rudder]. The same effect occurs in the pitching plane with the propwash bending in the appropriate direction to raise or lower the nose, acting like the elevator. Both actions stabilise the aircraft. Conversely, the tractor prop may aid the stability but is mainly de-stabilising. If a tractor prop aircraft is induced to yaw to the left (as in the example above) the propwash will deflect to the right pulling the nose further to the left, resulting in a de-stabilising moment. Again, the same de-stabilising factor is present in the pitching plane.

Another de-stabilising condition is a combination of up-elevator and an increase in engine power. Consider an aircraft in the flare, about to touch down with the engine throttled back to idle power and up-elevator applied to hold off. The pilot then decides to make a late go-around and applies full power. The increase of propwash over the elevators causes an increase in elevator effectiveness that in turn, causing the nose to pitch up further, the result is a de-stabilising motion. On some high performance

Aircraft from the WW I era used pusher props to leave the front end of the fuselage open for the gunner. This Vickers F.B.5 Gunbus is located in the RAF Hendon Museum, London.

single-engine aircraft, the engine and propeller are not aligned with the aircraft's axis but are tilted downwards two or three degrees and to the right by a like amount. The degree of nose down tilt depends on the aircraft's power loading (weight/power). When the aircraft is flying level with a nose-high attitude, as in the landing flare, the propwash inflow to the prop disc will be parallel to the direction of flight. However, on passing through the prop disc which is tilted rearwards, the propwash will be deflected downwards and in compliance with Newton's Third Law, this reacts on the prop disc as a nose pitch-up, combined with the thrust/drag vector, all part of the four forces acting on the aircraft. By tilting the engine and prop downwards, this places the prop disc closer to being at right angles to the propwash inflow, and therefore, reduces the downward deflection of the propwash and hence, the nose pitch-up. The outcome of this is an improvement in pitch stability.

Helical Propwash

As the propeller rotates, it produces thrust, drag and torque. It is the propeller's drag component that causes the propwash vortex sheet emanating from each prop blade to be whirled around on a helical path, or corkscrew fashion. This leads to a loss in propeller efficiency of just under two percent. The slipstream, due to the aircraft's motion through the air, will be flowing straight back over the fuselage. In effect, it can be considered as sliding over the tailplane unnoticed, whereas the helical vortex propwash will strike the tailplane in a series of pulsations. At a low air speed, the tail will experience more pulsations per unit time than at a high speed due to the vortex coils being closer together. This results in the propwash vortex sheets striking the fin and rudder at a greater angle of attack causing an increase in yaw to the left. At high air speeds the coils will be relatively elongated and the angle of attack on the tail and fin will be reduced resulting in less yaw. The rotating propwash will also strike the underside of the port main wing and stabiliser at an increased angle of attack causing an increase in lift. At the same time, the rotating propwash will strike the top surface of the starboard wing and stabiliser at a reduced angle of attack, resulting in less lift. The net result is a rolling moment, this time to the right, which under some conditions could counteract the induced yaw to the left, caused by the propwash striking the fin and rudder, thus aiding stability.

'P' Factor

The term P-factor (or in full, the propeller factor) is familiar to most pilots, although the term fails to explain the actual problem. The more descriptive term is asymmetric blade effect, associated to the blade element theory, which deals with the difference in thrust on the up-going and down-going prop blades. Closely related, is the asymmetric disc loading, which is associated to the axial momentum theory, and deals with the air mass flowing through the prop disc.

Due to the propeller axis being inclined to the direction of flight, one half of the propeller disc produces more thrust than the other half. Tail-dragger aircraft during take-off are more prone to P-factor than nose-wheel aircraft, due to the fact propeller thrust is greatest at high power settings and low air speed.

Assuming a two-blade propeller, P-factor is caused by the difference in angle of attack and the velocity between the up-going and down-going blades, with velocity being the major factor. In straight and level flight, the propeller axis

4 – Effect on the Aircraft's Stability

The Douglas EA-1F Skyraider with its powerful radial piston-engine driving a four-blade propeller is a prime candidate for P-factor. This Skyraider is located at the Pima Air & Space Museum, Tucson, Ar.

is parallel to the airflow through the propeller disc, and the angle of attack of each propeller blades remains the same. Increasing the angle of inclination of the propeller axis increases the difference in each blade's angle of attack up to a maximum inclination of 45°. Between 45° and 90° inclination, the difference in angle of attack reduces to zero degrees. A tail-wheel aircraft sitting on the ground has its prop axis inclined to the horizontal and the difference in angle of attack is relatively small at approximately ½° to 1°, producing a six percent difference in lift coefficient. The up-going and down-going blades meet the airflow with a difference in velocity of around seven percent. Because the velocity is squared in the thrust (lift) formula, it has greater effect than the angle of attack, and when the propeller axis inclined, the speed ratio of both blades is different, although the forward speed and RPM, are both constant.

To explain this anomaly, consider the propeller axis inclined 90° to the direction of airflow, as found on helicopter main rotor blades. The helicopter's advancing rotor blade has a constant rotational speed (RPM) plus the helicopter's forward speed. The retreating blade has the same constant RPM minus the helicopter's forward speed producing a difference in speed between the advancing and retreating blades. The aircraft propeller axis inclined just a few degrees, experiences a more subtle effect, but the velocity difference is still there. The down-going propeller blade having greater velocity and increased angle of attack produces more thrust on that half of the propeller disc, and is the major contributor to asymmetric disc loading, or P-factor.

Gyroscopic Effect

Any spinning mass, the propeller included will be affected by gyroscopic rigidity and precession. Rigidity is the tendency of the spinning mass to remain with its axis in a fixed position relative to space and to resist any force that tries to move it.

If an applied force succeeds in displacing the spinning mass, the resulting movement is known as gyroscopic precession, which acts as if the force was applied at a position 90° around the plane of rotation from the applied force. Tail-wheel aircraft are more prone to instability problems on take-off or landing than a nose-wheel aircraft due to the gyroscopic effect caused by the inclined prop shaft.

Consider a tail-wheel aircraft with a right-hand propeller, as the tail is raised during the take-off run, it acts

The Dornier Do-335 Pheil is a WW II centre-line thrust aircraft. Two Daimler-Benz V-12 piston engines of 1750 BHP each drive three-blade feathering props. This sole example is located in the Udvar-Hazy Centre, Chantilly, Virginia, USA.

Gyroscopic Effect

as if a force is being applied to the top rear of the propeller causing it to tilt forward. However, because of precession, the propeller acts as if the force had been applied to the rear right-hand rear side of the propeller disc, causing the aircraft to turn to the left. The gyroscope force affects all propeller-driven aircraft whenever the propeller axis is forced to tilt. An increase in propeller weight, RPM, diameter, and the rate of pitch, roll and yaw movements will cause greater gyroscopic effect, which can be quite noticeable during in-flight maneuvers. During a steep turn to the left, a right-handed propeller will cause the plane's nose to initially rise, and a steep turn to the right will cause the nose to drop. Therefore, pitching the nose up produces a right yaw, whilst a nose-down pitch will cause a left yaw. Hence, the need to apply rudder to maintain balance during maneuvers.

In a sideslip maneuver, 'P' factor causes a pitching moment and the nose may rise or drop depending on prop RPM, direction of rotation and direction of yaw.

Spinning maneuvers can be adversely affected due to the greater rate of pitch, roll and yaw. The nose attitude in the spin can be either flattened or pitched down further. This will depend on the aircraft type, the control surface movements and if the aircraft is in an inverted spin, or not. Other factors such as the aircraft weight, centre of gravity location, aerodynamic or centrifugal forces, can all affect the spin behavior. The pitch-up during a spin to the right will intensify the right yawing and rolling moments, resulting in a faster rate of spin with a steeper nose attitude. Conversely, a spin to the left will be flatter due to the nose being raised by the prop's precession force associated with left yaw. A flat spin with its greater angle of attack is always harder to recover from than a steep, nose-down spin. Getting the nose down to reduce the angle of

4 – Effect on the Aircraft's Stability

The Cessna 336/337 Skymaster has centreline mounted engines mounted at each end of the fuselage.

attack and throttling back the engine are prerequisites to spin recovery. Having the power on during the spin can cause adverse affects, due to the uneven alignment of the propwash flowing over the outer wing (due to yaw) creating greater lift and less drag than the inner wing, this will cause an increase in roll and yaw.

In conclusion, the effect on the aircraft's stability and its tendency to yaw to the left on take-off with a right-handed prop will depend on the propeller's torque, 'P' factor, prop location and gyroscopic effect. In addition to this left yaw, the propwash could cause an opposing rolling moment to the right. On some aircraft types where the above effects can be too great a problem, the aircraft may have a contra-rotating propeller installed to eliminate, or at least, reduce some of theses undesirable effects.

Centreline Thrust

Aircraft with centreline thrust have two piston-engines mounted in tandem on the aircraft's centreline. The loss of one engine alleviates the asymmetric thrust condition associated with twin-engine aircraft with wing-mounted engines. Cessna used the push/pull configuration on their twin-engine Cessna 336/337 Skymaster, with an engine mounted at each end of the fuselage pod. It is more commonly known as a centreline thrust configuration. The first aircraft with centreline thrust was designed and patented by Claudius Honoré Dornier (1884–1969) a German airplane designer and manufacturer. One of his more famous designs, in the 1940s was the Dornier Do 335 Pheil (Arrow) heavy fighter with two piston-engines mounted at each end of the fuselage in the centreline thrust configuration. The aircraft arrived too late in WW II to see active service.

Minimum Control Speed (V_{MC})

The minimum control speed (V_{MC}) is the speed at which a multi-engine aircraft can fly with a failed engine and still maintain directional control.

The V_{MC} speed is determined by the force from the rudder required to maintain directional control to counteract the yaw force caused by an engine failure. Below the V_{MC} speed, rudder authority is reduced and the aircraft will yaw and diverge from the required heading. It was mentioned above: a twin-engine aircraft with both propellers rotating in the same direction has the greatest yaw force with the critical engine failed. To be precise, there are two different air speeds at which the rudder fails to maintain directional control: there is a V_{MC} for each engine. However, the higher of the two air speeds is taken as the operational V_{MC}. Aircraft with counter-rotating propellers have the same amount of yaw force with either engine failed, therefore, the V_{MC} is the same when either engine is failed.

4 – Effect on the Aircraft's Stability

The term V_{MCA} applies to the minimum engine failure control speed when the aircraft is airborne. The V_{MCA} should be no higher than 1.2 times the stalling speed. The term V_{MCG} defines the minimum control speed on the ground, and it must be lower than the take-off decision speed (V_1) to ensure directional control can be maintained following an engine failure.

Counter-rotating Propellers

On twin-engine aircraft, the props of each engine may rotate in opposite directions with the top blade rotating in towards the fuselage. These are known as counter-rotating propellers.

The main advantage of counter-rotating propellers is during take-off and climb-out after an engine failure. On a conventional twin-engine aircraft with both propellers turning clock-wise, asymmetric thrust causes the greatest yaw when the left-hand engine is shutdown. This is due to

A front-end view of a Lockheed P-38 Lightning showing its counter-rotating propellers. This aircraft is located in the National Museum of the USAF, Dayton, Ohio.

COUNTER-ROTATING PROPELLERS

thrust generated on the down-going side of the propeller disc, remember the P-factor or asymmetric disc loading! On right-handed propellers the center of thrust to displaced to the right of the propeller axis. On the right-hand engine it is further away from aircraft's normal axis, and the centre of thrust on the left-hand engine will be closer to the aircraft's normal axis. If the left-hand engine fails, the right-hand engine will produce the greatest yawing moment due to the centre of thrust being displaced further outboard. The drag of the windmilling left-hand prop will contribute to the yawing force. In this instance, the left-hand engine is said to be the critical engine, due to the greater yaw force caused by the thrust from the right-hand engine.

On a twin-engine aircraft with counter-rotating props, both props will have the centre of thrust an equal distance from the aircraft's normal axis. Therefore, the failure of either engine will produce an equal yaw force. The critical engine is eliminated and single-engine performance will be the same with either engine failed.

Airplanes with propellers rotating anti-clockwise, or 'left-handed' propellers, will have their right-hand engine as their critical engine. The 'critical' engine is so named due to the control problems being more critical when the critical engine is shut down. The Fokker F.27 Friendship is one aircraft that comes to mind with a right-hand (Number 2) critical engine, due to the left-hand rotation of its propellers powered by their Rolls Royce Dart fixed-shaft turboprop engines.

The location of the wing-mounted engines on twin-engine aircraft is also important. Placing the engines too near to the fuselage will not only increase noise in the passenger cabin, it can also affect the amount of thrust produced by the propeller. The closeness of the fuselage affects the free air flow between the prop and fuselage.

Four sets of contra-props power the Avro Shackleton AEW.2 maritime patrol aircraft. This aircraft resides in the Museum of Science & Industry, Manchester, England.

This has an affect on the prop by slightly reducing the prop thrust of the prop on one side of the aircraft, while the prop on the other side remains unaffected. Although this imbalance of thrust is not as great as the 'P' factor or asymmetric disc loading, it is still present to a certain degree.

Contra-rotating Propellers

Two co-axial mounted propellers driven by the same engine, but rotating in opposite directions are known as contra-props. Using two propellers mounted on the same co-axial shaft with a given propeller diameter, will absorb

The Fisher P-75A Eagle, the last of 14 built, has a contra-prop powered by an Allison V4320 engine of 2600 BHP mounted amid-ship. It must rank as one of the earliest contra-prop aircraft to be built, (in 1943). This aircraft is displayed in the Research Section of the National Museum of the USAF, Dayton, Ohio.

4 – Effect on the Aircraft's Stability

a greater amount of horsepower than a single prop unit. The rear-mounted propeller in the pair straightens out the helical propwash from the front propeller, which reduces the total propeller torque to zero and hence, take-off yaw and in-flight yaw caused by power changes. This is the important factor on high-powered aircraft.

Additionally, the wing's structural loading on multi-engine aircraft will be greatly reduced due to the absence of prop torque. On the down side, the disadvantages are the increased weight and complexity of the co-axial prop shafts. Contra-props have their own distinctive noise due to the rear prop interrupting and reacting on the helical propwash vortex formed by the front prop.

Contra-props mounted on the Fairey Gannett AEW3 in the Yorkshire Air Museum, Elvington, England.

5 – Prop Tip Speed & Noise

Tip Speed & Noise

Have you ever had the privilege to hear the beautiful sound of a Merlin-powered aircraft take-off or fly overhead? Or heard the rasping sound of a rowdy radial engine, or the thundering roar of a jet aircraft taking-off? All this noise is sweet music to a pilot's ears, but not so for local residents living near an airport. To the locals, it is a very disturbing nuisance. To this end, aircraft and propeller manufacturers all attempt to reduce aircraft noise as much as possible. In fact, certification requirements for all new aircraft designs stipulate the maximum allowable noise limits.

The Cause of Noise

Noise is generated by the engine, exhaust system, propeller propwash and the prop itself. The prop noise is dependant on the blade loading, number of blades, prop diameter, and the location of the prop on the aircraft. However, the main cause of noise is the propeller tip speed.

Consider a two-blade prop installation: the prop produces an inherent vibration once per revolution that will vibrate through the airframe to be heard as noise. The greater the number of blades, the less is the vibration and noise produced. Single-engine aircraft have their prop wake striking the cockpit windshield adding to the vibration and noise as opposed to multi-engine aircraft with their props further away from the cabin.

Tractor props mounted on the aircraft nose or in front of the wings are generally quieter than pusher props, which operate in the disturbed air flow passing over the aircraft creating resonance or noise in the cabin. However, the greatest amount of noise is heard the prop's plane of rotation. On a twin, or multi-engine aircraft, any occupants seated in line with the props will suffer the most noise. Moving the engine/prop further out board on the wing will help to reduce the noise heard in the cabin, but this will also increase engine-out asymmetric forces, as mentioned above.

'Shrouded props', or 'Propulsors' are claimed to be considerably quieter than conventional props due to the shroud around the propeller and also the lower tip speed (they are usually props of smaller diameter). But, the disadvantage here is, the prop noise can be directed more fore and aft by the shroud. Therefore the amount of noise heard to a certain extent is dependant on one's external position relative to the aircraft, or their position inside the aircraft.

High-speed Aerodynamics

Moving into the area of high-speed aerodynamics as applied to propellers, the definitions of the speed of sound and its associated critical Mach number and the effects of compressibility will now be considered.

The speed of sound varies with the ambient temperature and air density. Because air density is closely related to temperature, it can be ignore in the calculations. At sea-level where the International Standard Atmosphere (ISA) temperature is assumed to be +15 °C (288 K) the speed of sound as 661 knots and varies in proportion to the square root of the Absolute temperature. The speed reduces to 575 knots in the Stratosphere where the temperature is assumed to be at minus 56.5 °C (or 216.65 K). The speed of sound is also known as acoustic velocity, Mach number 1, or more simply as Mach 1, after Dr. Ernst Mach (1838–1916) the Austrian philosopher and physicist. Mach number is the ratio of aircraft speed to the speed of sound, or in this case the propeller tip speed to the speed of sound.

At subsonic speeds, below Mach 0.8, air acts as if it is incompressible and this assumption is fine until speeds greater than 300 knots and altitudes of 10,000 feet are considered and where the effects of compressibility and air density can no longer be ignored. The aircraft wings, or prop blades, cause compressibility as they move through the air sending pressure waves ahead of it, which travel at the speed of sound and cause the approaching air flow to separate and travel over and under the wings or prop blade surfaces. As the prop tips approach the speed of sound, the pressure waves have less time to move ahead of the blade and they eventually become stationary on the blade's surface. This causes an increase in compressibility resulting in a serious loss in propeller efficiency, which in turn, reduces thrust and causes a rapid rise in blade drag, prop torque and especially noise.

Altitudes above 25,000 to 30,000 feet are the domain of jet propelled aircraft and not many propeller-powered aircraft are found there, due to the reduction in air density, air temperature and pressure. Propellers are unable to cope with the reduced air density at these high altitudes and at 40,000 feet the thrust produced by the propeller is reduced to around 25% of the sea-level value. How does the reduced air density at high altitudes affect the thrust produced by the prop? An inspection of the familiar lift formula below shows it can be applied to the propeller thrust:

Thrust (or lift) = $C_L \frac{1}{2} \rho V^2 S$

5 – Prop Tip Speed & Noise

Where C_L = lift coefficient
½ = a constant
ρ = air density
V = prop RPM in FPS
S = prop blade area

If the blades area (S) is assumed to be constant and the air density (ρ) is reduced at altitude, the lift coefficient (C_L) and prop speed (V) are the two variables. The lift coefficient can be increased by increasing the blade pitch angle but this increase is offset by the prop speed decreasing. [Remember that coarse pitch produces lower RPM]. Therefore, with the lift coefficient, RPM^2 and blade area all doing nothing to increase prop thrust, it follows the remaining factor of air density is the only remaining variable and because air density decreases with altitude, prop thrust must also decrease.

It is common knowledge air temperature decreases along with air density as altitude increases. Not only is the prop's thrust decreasing with altitude but the speed of sound is also decreasing. This has a detrimental affect on propeller performance. The prop blades are working closer to the speed of sound at altitude than they do at sea-level due to the difference in air temperature. This can be shown by calculating the tip Mach number (Mt) for a given aircraft's prop at sea-level (s/l) and for example 20,000 feet from the formula:

Tip Mach number = V. tip/speed of sound

Where: Prop tip speed (V. tip = 870 FPS
S/l speed of sound = 1100 FPS
20,000 feet speed of sound = 1040 FPS

Mt at s/l (15 °C) = 870/1100 = Mach 0.79
Mt at 20,000 feet (−25 °C) = 870/1040 = Mach 0.83

In the above figures, the prop has a higher tip Mach number at altitude and it is therefore operating closer to the speed of sound with efficiency deteriorating. Closely associated to the speed of sound is the term critical Mach number. This is the speed at which the airflow over a body, or prop blade, reaches Mach 1.0, due to the blade's curved upper surface, while the prop blade itself is actually moving at a speed below Mach 1.0. It is the thickness/chord ratio that determines the critical Mach number of an airfoil, which increases with thinner prop blades with a high aspect ratio and is therefore more important operationally than the speed of sound.

Incidentally, when the prop tip approaches the speed of sound, a condition occurs known as 'cavitation', caused by a near vacuum on the suction face of each prop blade near the tips, which again reduces efficiency.

Tip Speed

The prop tip speed can be found, in feet per second, given the RPM and prop diameter from the following formula:

Prop tip speed = $2\pi RN$

Where: 2 = a constant
π = 3.14...
R = prop radius in feet
N = prop revs per second

Given the following figures the prop RPM in FPS can be found:

Prop radius = 74 inch/2 = 37 inch = 3.08 feet
Revs per second = 2700/60 = 45 revs per second

5 – Prop Tip Speed & Noise

$$\begin{aligned}\text{Prop tip speed} &= 2\pi RN \\ &= 2 \times 3.14... \times 3.08 \times 45 \\ &= 870 \text{ FPS}\end{aligned}$$

An alternative and simpler formula to the one above is as follows:

$$\begin{aligned}\text{Prop tip speed in FPS} &= \frac{\text{RPM} \times \text{diameter}}{229.3} \\ &= \frac{2700 \times 74 \text{ inches}}{229.3} \\ &= 870 \text{ FPS}\end{aligned}$$

In the example above, the propeller rotating at 870 FPS on take-off at full RPM would be operating very close to Mach 0.8, where efficiency begins to deteriorate and the prop's noise level is about to exceed the maximum allowable limits. An aircraft designer may consider a propeller reduction gear to reduce the tip speed to a value of 0.7 to 0.8 of the engine speed. One example of using prop reduction gear was demonstrated on the Australian CAC Wirraway, a licensed built version of the North American Harvard. The Wirraway's geared engine driving a three-blade prop at a slower RPM was much quieter than the Harvard's distinctive growl on take-off.

So far, we have only considered the effects of the rotational velocity on the prop tip speed (vector A–B on Diagram 2, Propeller Terminology). To this vector we must add the vector representing the props advance per rev (B–C). We now have the third vector in the triangle (vector A–C) corresponding to the propeller's helical flight path. Because the vector A–C is of greater length than vector A–B, it follows the prop tip speed will be higher when the aircraft's forward speed is increased from zero up to cruising speed. Using the above formula again, and the 74 inch propeller turning this time at 2400 RPM, we find the prop tip speed to be approximately 775 FPS (236 m/sec) when the aircraft is stationary. Increasing the plane's forward speed up to 142 knots, the prop tip speed increases to around 810 FPS (247 m/sec). The noise level would be fairly high but acceptable at this speed.

Generally at tips speed of around 600 FPS (183 m/sec) the prop will be relatively quiet but, the noise level will start to increase around 700 FPS (213 m/sec). At 880 FPS (268 m/sec) the prop could be unacceptably noisy, as mentioned above. When the prop tip speed approaches the speed of sound, compressibility problems and tip vortex losses increase, which in turn reduces thrust, efficiency and increases prop torque and noise. Mach 0.8 or 880 FPS (268 m/sec) is about the maximum tip speed a normal prop can safely operate to, but there are a few exceptions with some props designed to run at transonic tip speeds (Mach 0.8 to Mach 1.20).

Tips, Blade Shape & Materials

Wooden propellers due to their inherent thickness required for structural strength, are more suitable for low speed aircraft with lower tip speeds operating up to 660 FPS (200 m/sec) maximum. At higher speeds, the thicker sections create more drag and compressibility may become a factor. Metal props have thinner blade sections and they can operate to higher tip speeds: 880 FPS (268 m/sec) is an average maximum. New types of propellers now being produced have supercritical airfoil sections that can operate with high critical Mach numbers. They are less affected by the drag rise and noise associated with high tip speed due to their favourable thickness/chord ratios. Composite materials such as Kevlar, graphite, carbon or

5 – Prop Tip Speed & Noise

A Fokker E-III Eindecker of WW I era with a carved, laminated and scimitar shaped propeller. This aircraft is located in the Air & Space Museum, San Diego, Calif.

glass fibre are all used in the manufacturing of propellers. Composite materials are considerably stronger and lighter than wood or metal props, which helps to reduce the load on the prop hub. They also have a higher aspect ratio and operate to higher tip speeds than either wood or metal props, while still maintaining excellent efficiency.

Although wood propellers are usually associated with low powered aircraft, the German prop manufacturer MT-Propellers has been making props ranging from fixed-pitch props for light, home-built aircraft and right up to include constant-speed props with electric or hydraulic constant-speed units for powerful turboprops of up to 1500 shaft horsepower. The propellers, from two to six-blades per prop are made with a wood base covered with fiberglass. Their advantage is their lightweight, and also the fiberglass can be replaced if damaged.

Another factor that contributes to tip speed noise is the shape of the prop blade and its tip. For some reason, square tips tend to run quieter than round tips and from the early 1960s the square tip has become the one mostly used, with a few exceptions. The Piper Cheyenne III, introduced in 1979, was the first production aircraft to sport 'Q-tip' propellers as standard equipment. 'Q-tip' propellers have a form of end plate where the last two inches (5 cm) of the prop tip is bent upwards at a 90° angle, similar to winglets found on modern jet transport aircraft. The 'Q-tips' are there to reduce back-side pressure from leaking around the tips and to enable high blade span loadings to be achieved at a lower RPM than normal, thus reducing prop noise and improving efficiency. The advantages of 'Q-tips' is debatable and can vary between different installations.

The Aerospatiale ATR-72 turbo prop aircraft have modern six-blade scimitar shaped props.

5 – Prop Tip Speed & Noise

One advantage is the reduced prop diameter allows greater tip clearance from the fuselage on twin-engine aircraft, which will reduce cabin noise. The 'Q-tip' may alter the prop tip vortex reducing the amount of gravel picked up during take-off and hence reducing prop blade leading edge damage. Also, at low air speed and high power such as on the climb, is where they work the best.

Another modification for propellers was the introduction of the use of sweptback tips. They work in the same manner as swept back wings on a jet aircraft by increasing the blade's critical Mach number, allowing the prop to run at speeds closer to Mach 1.0, with reduced noise. Curved or scimitar shaped blades are an alternative to swept-back tips having a similar effect. The scimitar shape is ideal for Propfans, which will be covered later. They are also becoming more popular on new generation turboprop aircraft such as the Aerospatiale ATR commuter aircraft. The scimitar shape obviously provides better aerodynamic performance on today's modern aircraft, but the shape is not a new idea. Several early aircraft from the World War one era used scimitar shaped props. Was it for aerodynamic or aesthetic reasons the early prop designers chose the scimitar shape? After the WW I era, the style faded away and prop blades were made straight (usually with round tips) and now the scimitar shape is becoming more common again. During the 1960s, the slender or elliptical tip became more popular, possibly because elliptical tips have greater efficiency than square or rounded tips. The tip shape also has an effect on the prop's vibration characteristics – another problem for the prop designer to contend with. Hartzell and McCauley produce scimitar shaped blades for light aircraft.

The North American B-25J Mitchell bomber has rounded prop tips, typical of many aircraft until the 1960s when square tips became common. Pima Air & Space Museum, Tucson, Ar, is home to this aircraft.

5 – Prop Tip Speed & Noise

Synchronizing

One source of noise audible to the occupants of twin-engine aircraft occurs when the propellers are not synchronized. The noise can be heard as a throbbing sound when the props are 'out of synch' caused by a vibration due to the difference in RPM between each engine. When the props are correctly synchronized the noise becomes a steady hum. The vibratory frequency is only a discomfort to the occupants and does not in any way affect the structural integrity of the airframe.

The propellers can be synchronized in one of four different methods. The first two manual methods apply to light twin-engine aircraft, and the following two methods are applicable to larger more sophisticated aircraft. On a twin-engine aircraft with fixed-pitch props (not many around these days) the required RPM/power is set first with the throttles and then one throttle is slowly adjusted either way until the throbbing noise becomes a steady hum. With constant-speed props, the required manifold pressure is selected with the throttles followed by selecting the chosen RPM with the prop pitch levers in the usual manner. One prop lever is then adjusted either way to synchronize the props to a steady hum. Depending on the make of engine, some props are more easily synchronized by retarding the prop pitch lever, and on other engines the lever should be advanced. The throbbing beat will slow down and merge into a steady hum when the control is moved in the correct direction, while a quickening of the beat indicates the control is being moved the wrong way, so readjust accordingly.

Turboprop aircraft employ the use of a prop synchronizer. This is an electrical system comprising a generator mounted on each engine, with one engine (usually the right-hand engine) being the master unit.

Signals from the left-hand engine's generator are adjusted electrically to match those of the master unit, ensuring all propellers rotate at the same RPM.

The fourth method uses a prop synchrophaser, developed by Hamilton Standard in 1978. The synchrophaser controls the engine RPM exactly by ensuring the prop blades on each engine pass through the same angular location at the same time. The prop's RPM is fine-tuned electrically by the phase signals from each engine's generator. When the props are turning with a phase difference of plus or minus one degree, a noise reduction of to 6.5 decibels (6.5 dB) is possible, while a phase difference of plus or minus five degrees produces a 4.5 dB reduction. Commuter and business aircraft usually employ the ± 5° phase due to the lower installation costs. A control is provided for the pilot to fine tune the system by altering the phase angle slightly in order to obtain the preferred sound level. The system can also be turned on or off as required, but must always be turned off for take-off and landing in case of engine failure. Decreasing engine RPM on the failed engine would cause the other engine to follow suit, if engaged.

How Noisy are They?

So far, only a brief reference has been made to the amount of noise produced by the props. So, how noisy are they? Answer – quite a lot, but to be more precise... the noise will be in the region of 76–80 dB during take-off for a high performance single-engine aircraft, decreasing to around 70 dB or even lower for an aircraft with a fixed-pitch propeller. Fixed-pitch props are generally quieter due to the prop's relatively smaller diameter and lower RPM, which are the two main factors that influence the tip speed and prop noise. It follows, a fixed-pitch prop with a relatively

5 – PROP TIP SPEED & NOISE

A Lockheed YO-3 Quiet Star observation aircraft, with a three-blade, wide chord prop. This is one of the quietest planes built due to the very quiet propeller and other features. It is housed in the Pima Air & Space Museum, Tucson, Ar.

short diameter does not attain full RPM during take-off and because its tip speed is typically around 600 feet per second at 2500 RPM, it will be relatively quiet. Conversely, a high-performance single with a constant-speed prop will attain maximum RPM (2700–2900) during take-off and climb out, and will generate considerably more noise due to the greater RPM and higher tip speed. For any given aircraft, a three-blade prop will usually be of shorter radius than a two-blade propeller and three-blades are preferable because they generate more acceptable sound frequencies than two-blade props.

The American FAA 36 noise regulations (1988) places a limit of 80 dB maximum for the certification of new aircraft types, but some older aircraft can easily exceed this figure. It is up to the individual pilot to operate his/her aircraft as quietly as possible to avoid undue noise to airport neighbors. A noise abatement departure can be achieved by reducing the RPM as soon as is safely convenient after take-off and then climbing at the maximum rate of climb to gain as much altitude as possible before reverting to the cruise/climb procedure. The intensity of the plane's noise will decrease inversely with the square of the distance from the plane, altitude and distance are both excellent buffers of noise. However, what noise is sweeter than that of an aircraft taking-off and flying overhead? It sure beats 'heavy metal' rock music!

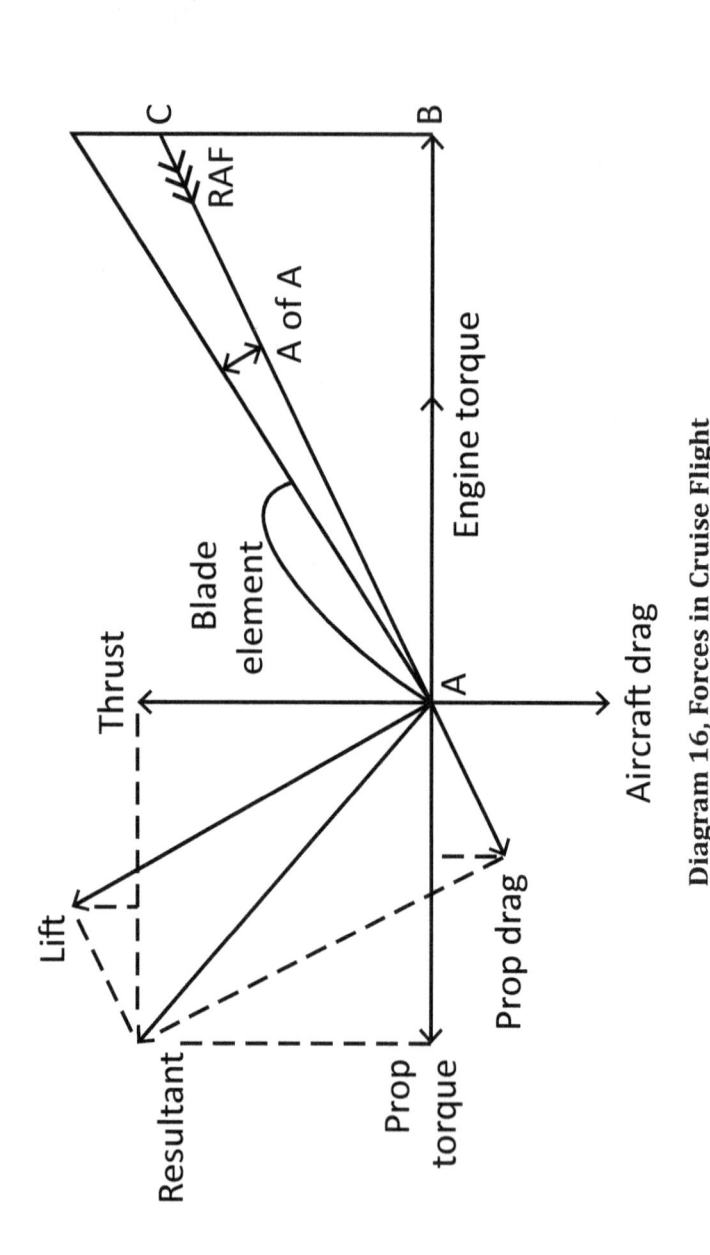

Diagram 16, Forces in Cruise Flight

6 – Propeller Forces & Stress

Prop Forces in Cruise Flight

With reference to Diagram 16, Forces in Cruise Flight, which is an extension of Diagram 2, Propeller Terminology, shows the propeller's rotational velocity (A–B) in the prop's plane of rotation and the aircraft's forward velocity (B–C). The vector (A–C) represents the relative air flow (marked RAF) or the air inflow velocity into the prop disc (from C to A). This is determined by the prop's speed ratio (the ratio of the aircraft's forward speed to the prop RPM). The vector A–C also represents the helical flight path of the blade as it travels in the opposite direction, from A to C.

Due to the prop blade's motion along the helical path and its angle of attack, an aerodynamic resultant force is produced, the same as found on the wing in flight. The resultant force can be divided into the components of lift and drag, but of greater importance, into the components of thrust and prop torque. The blade's lift coefficient, air density, and inflow velocity and blade area govern the strength of the resultant force. This is shown by the familiar lift formula, Lift (or thrust) = $C_L \frac{1}{2}\rho V^2 S$.

The propeller torque acts in the opposite direction and is equal to the engine torque at constant RPM. Ignoring the effects of a constant speed unit for now, the RPM will remain constant as long as these two forces are equal and opposite. Also, the prop's thrust is equal and opposite the aircraft's total drag. The forward component of the resultant force (the dotted line under the word 'lift') is equal and opposite the rearward component of prop drag (shown by the dotted line above the words 'prop drag'). These two forces cancel out leaving prop thrust to equal prop drag.

6 – Propeller Forces & Stress

The vectors of relative air flow, thrust and torque, etc, have been drawn as emanating from the trailing edge of the prop blade element. In practice, this is not true because the forces all act from the blade's centre of pressure. This diagram and the following two have been drawn this way for the purpose of clarity. Note here, how the blade element theory differs from the axial momentum theory covered in the section on Propwash-Thrust.

Windmilling Prop Forces

In the event of an engine failure, the prop will continue to turn or windmill for some time after the engine has failed if the prop is not feathered. The force required to turn the prop is provided by the energy in the air flowing through the prop disc due to the forward motion of the aircraft. Referring to Diagram 17, Windmilling Prop Forces, the propeller's rotational velocity (A–B) is assumed to have decreased but the aircraft's forward speed is still relatively high (assume the aircraft to be in a glide). The vector (A–C) or relative air flow (RAF) will strike the prop blade at a negative angle of attack assuming the prop path to be unchanged for the moment. In this condition, the helix angle will be greater than the blade angle and the angle of attack will be negative, the reverse of the normal cruise positions. The negative prop torque acts in the direction of propeller rotation powering the prop to a fast idle speed. With the engine failed, there is no engine torque to oppose prop torque. If the forward speed is increased by lowering the nose, the advance per rev vector (B–C) will be extended, increasing the negative angle of attack and in turn, increase the prop RPM.

By reducing the aircraft's forward speed, the windmilling prop drag can be reduced. Less drag is always an advantage

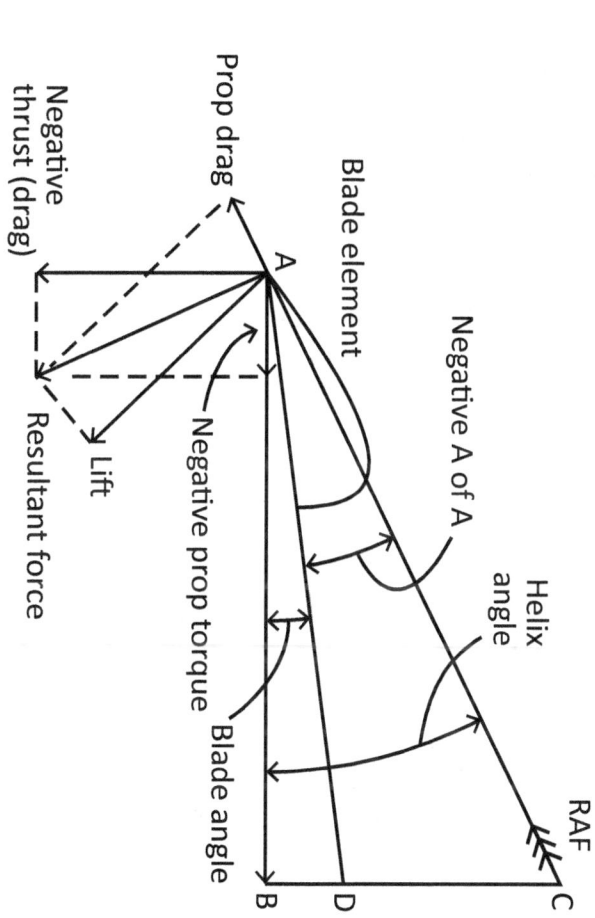

Diagram 17, Windmilling Prop Forces

6 – Propeller Forces & Stress

on single-engine aircraft because it reduces the angle of descent allowing more time and a greater distance to be covered in the ensuing forced landing. The reduced windmilling drag on a twin-engine aircraft will lower the minimum control speed (V_{MC}) enhancing engine-out handling. Diagram 17, also shows a reduced forward speed will reduce the advance per rev vector (B–C). The helix angle (AB–AC) will also reduce, thereby reducing the prop blade's negative angle of attack, which in turn reduces the lift coefficient and the resultant force and hence, it reduces the negative thrust (or drag).

The resultant force acting in the rearward direction produces undesirable negative thrust, known as parasite drag, or more commonly as windmilling drag. The negative prop torque component of the resultant force acts in the plane of rotation causing the prop to windmill. The

The propellers on this Boeing KC-97G Stratofreighter are all parked in the feathered position; this is unusual for a piston-engine aircraft. The aircraft is stored and displayed at the Pima Air & Space Museum, Tucson, Ar.

amount of drag produced depends on the pitch of the prop blades. Additional drag is produced when the prop is set to fine/flat pitch due to it greater negative angle of attack. A coarse pitch setting should be selected for a lower blade angle of attack, less drag and a lower critical speed on a twin-engine aircraft. Coarse pitch will also help to stop the prop from rotating. The windmilling prop drag reduces the overall lift/drag ratio of the aircraft from an average 15:1 to 11:1. The result will be a much steeper glide angle, hence the need to select coarse pitch. The drag of a turboprop's windmilling propeller can be excessively high and dangerous, with the fixed-shaft turbine engine being the worst case due to its inherent design. The drag of a free-turbine engine is only 25% of the fixed-shaft engine.

Reverse Thrust Forces

Reverse (or negative) thrust is achieved by turning the prop blades about 30 degrees past the fine/flat pitch stop to a negative angle of attack, referred to as braking pitch. In this position, the resultant force is acting in a rearwards direction, the same as when the prop is windmilling. However, during reverse thrust operations, prop torque is acting in the opposite direction to prop rotation opposing the engine torque as shown by Diagram 18, Reverse Thrust Forces. Engine torque is equal and opposite to prop torque acting in the direction from A–B as indicated by the arrowhead. This could cause the prop to attempt to windmill backwards against the power of the engine. The rearward component of the resultant force produces negative thrust to provide a very effective air brake, assisted by the quite considerable prop drag. With the aircraft reducing speed during the landing roll, the vector B–C will reduce. This will cause the negative angle of attack

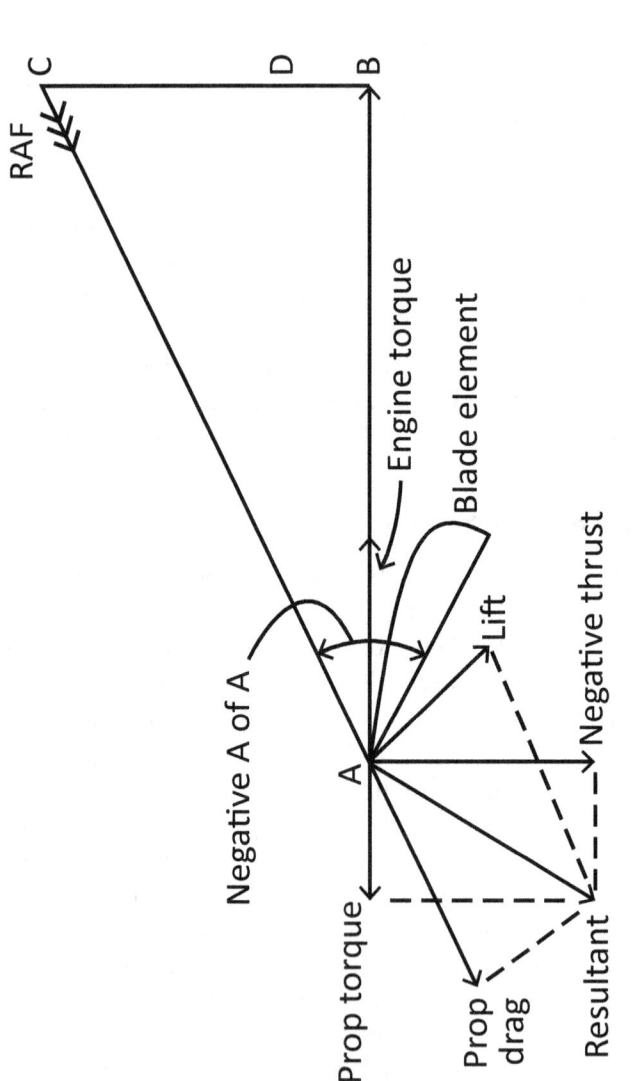

Diagram 18, Reverse Thrust Forces

to also reduce, resulting in a decrease in negative thrust. Hence the need to apply reverse thrust early in the landing roll to take advantage of its greatest effect. When operating in the reverse mode, the prop blades are at a negative angle of attack and the prop will be less efficient than at normal forward motion pitch settings. However, full efficiency is not essential due to the brief period reverse is used on the landing roll.

Prop Stress

Material stress can be divided into four main categories of torsional, tensile, compression, and bending stress. Any one stress or combinations of stress can be present acting on an aircraft or propeller at any one time. Stress is measured as a force per unit area. The stress will vary depending on the operating conditions of the propeller, the stress force can be reversed if the propeller is windmilling or with reverse thrust application. Life for a propeller is not an easy one!

Reference to Diagram 19, Blade Stress Forces, shows the centrifugal force caused by the spinning propeller acting from the propeller hub through the leading and trailing edges of the propeller blade. The centrifugal force is divided into two components, one causing tension (tensile stress) due to stretching and the other component causing a centrifugal turning moment (torsional stress).

The tensile stress tends to flow through the propeller blades outwards, in effect stretching the blade in length and generating a considerable force on the propeller hub and increases towards the propeller tip.

The centrifugal turning moment (torsional stress) acting at 90 degrees to the propeller axis produces a turning moment around the propeller pitch change axis

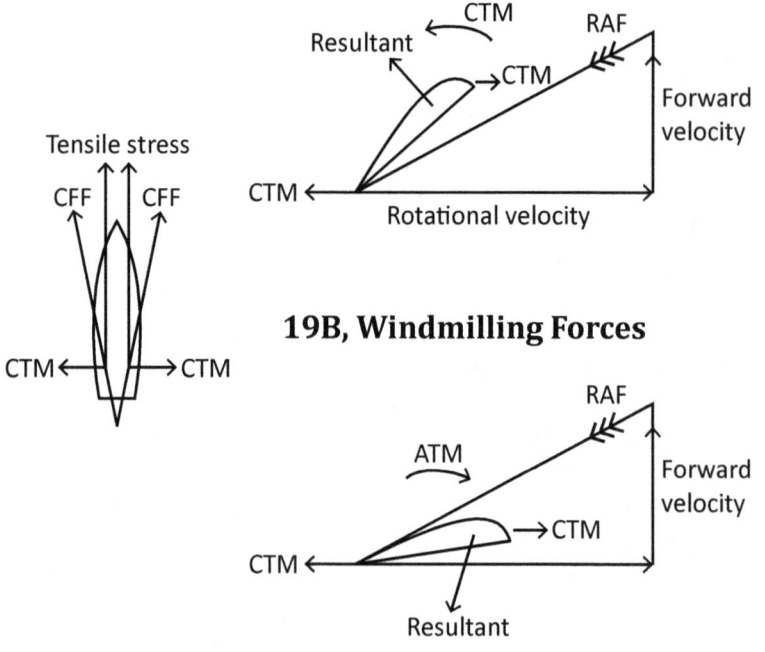

Diagram 19, Blade Stress Forces

tending to turn the blades towards fine/flat pitch. This force is opposed by the pitch change mechanism, which is placed under stress, but is assisted to a certain degree by the aerodynamic turning moment during normal cruise conditions, which tend to turn the blades towards coarse pitch. The pitch change axis is normally to the rear of the blade's centre of pressure. The resultant aerodynamic force acting at the centre of pressure will tend to turn the propeller blades in the opposite direction; the direction is reversed when the propeller is windmilling causing the resultant force to act rearwards. The aerodynamic turning moment is acting in the same direction as the centrifugal turning moment and turns the blades towards fine/flat pitch. The centrifugal turning moment is a more powerful force than the aerodynamic turning moment and causes the most stress. Torsional stress forces increase with the RPM squared.

Additional to the stress mentioned above, is the bending stress on the blades caused by the thrust (blade loading) of the propeller. The blade loading is measured in horsepower per square foot. The force generated by the propeller tends to bend the blades forward when under power, while the tension in the blade caused by the centrifugal force opposes the thrust, which tends to straighten the blades out. Bending stress will be reversed in direction with reverse thrust application on the landing rollout.

At normal cruise RPM, the stress on the prop blades will be approximately 4500 pounds per square inch. On some engine/prop combinations, a destructive vibratory frequency could exist at around 1900–2200 RPM. This torsional vibration can increase stress on the prop tips to that double experienced at normal cruise RPM. This is due to varying aerodynamic loads acting on the prop caused by

6 – Propeller Forces & Stress

a high vibration frequency coupled with the high inherent vibrations of the engine cylinder firing pulsations. If the vibration from the propeller coincides with those caused by the engine, a severe vibration will result. This is a different vibration to that caused by an out of balance prop and the pilot is unlikely to notice it. When the propeller's and engine's vibration synchronizes the amplitude of vibration can rapidly increase causing severe stress with possible blade failure or hub damage with disastrous results. A caution arc on the engine tachometer covers the RPM range to be avoided for continuous operation. This problem is only peculiar to certain aircraft: those aircraft with composite props are less prone to this vibration, while turboprops are totally free due to their lack if cylinder vibration. The red line at the top end of the tachometer scale indicates the maximum RPM permissible and exceeding the redline limit, apart from engine operating limitations, would allow the prop tips to reach the speed of sound. Additional stress would then be caused due to vibration flutter caused by the high propeller tip speed. Although safety margins are built in, red line limits should never be exceeded intentionally at any time.

Another form of stress can be induced by stone or gravel ingestion through the prop disc, and striking the blade's leading edge. This will cause damage in the form of knicks, or stress raisers, which can lead to stress concentrations. The blades flexing under normal loads, especially in the area from 12.5 cm to 23 cm (5 to 9 inches) from the tips can lead to metal fatigue and loss of the prop tip. The result will be extreme vibrations caused by the prop being out of balance and could lead to the engine being torn from its mounts. With the engine heading earthwards and the aircraft's centre of gravity moved well aft, it would be very difficult to achieve a balanced glide, the end result

The 1917, WW I, Bristol Bulldog IIA has prop tipping on the prop's leading edges. It is displayed in the RAF Hendon Museum, London.

can easily be imagined! If the engine is still onboard, but shaking badly, it is imperative the engine is shut down and the prop feathered if possible. If the propeller is not the feathering type, reduce air speed as low as possible until the engine stops. Prevention is the best cure; any knicks found on the prop should be filed, or dressed out to prevent a crack from forming. Avoid taxiing over loose gravel if possible, but if it is unavoidable reduce power to a low RPM. Never run-up an engine over loose gravel and when taking-off, open the throttle slowly to get the aircraft moving before going to full power. This technique will allow the prop to blow the gravel well clear without causing any damage to the prop blades. Tail-wheel aircraft have a greater clearance between the prop tips and the ground and are therefore less prone to stone damage than their trike counterparts.

Sharp taxi turns can induce stress on the prop because the prop acts in the same way as a gyroscope due to its spinning mass and will attempt to resist any changes in direction. All power changes should be made smoothly and gently to avoid any undue stress being applied to the engine or the prop attachment bolts. On a wooden prop, check the metal sheathing, (also known as prop tipping) on the leading edge of each blade is secure and free of any damage. Also, there should be no splits or cracks in the wood or separation between the wood laminations. A wood prop not used for long periods should be parked in the horizontal position to equalize the moisture content in each blade.

The ultimate stress force a prop can endure occurs when the prop blades strike the ground during a crash or wheels-up landing. Metal blades will bend either backwards or forwards during the impact. The resulting force, which in turn depends on a number of factors determine which way

PROP STRESS

The wooden prop blades on this Handley-Page Halifax II bomber were sheared off during its war time crash-landing. The Halifax is displayed as recovered in the RAF Hendon Museum, London.

they bend. These factors include the blade chord, shape, and pitch, and the aircraft's forward speed at the time of impact and also if the engine is providing power to the prop, or just idling. If the aircraft slide sideways the blades will buckle as well as bend. This kind of treatment is to be avoided to say the least! Wooden props and their engines can generally fare better in a crash landing. The wooden prop is more likely to break up than a metal prop and absorb some of the impact stress, protecting the engine somewhat. If you have the misfortune to forget to lower the undercarriage on the approach, you may hear the prop making ground contact as you flare for the landing. Do

6 – Propeller Forces & Stress

not attempt a go-around at this late stage in the hope of landing wheels-down on the next attempt. The damage is now done to the prop and the inherent forces associated with a full power climb could lead to total prop destruction and loss of the aircraft. It is safer to carry on with the landing and ride out the ensuing belly-ride. It must now be obvious the prop is a very important part of the aircraft and deserves to be treated gently with the utmost respect.

This Messerschmitt Bf 109e received prop damage in the crash-landing after being shot down in England, in 1940. This aircraft is on display in the Imperial war Museum, Duxford, UK.

The Balanced Prop

The propeller must be in static and dynamic balance to reduce stress caused by vibrations and make for a smooth running propeller. Generally, a three-blade propeller runs more smoothly than a two-blade prop; no matter how many blades there are it is important the propeller is in perfect balance.

If a two-blade prop has a 100-gram weight placed at the tip of one blade and 200-gram weight placed half way along the other blade, the prop would be in a state of static balance. A prop that is statically stable would not rotate on its prop shaft (if it were free to do so) from any position. The prop is then said to be in a state of static balance but it would not be dynamically balanced and would vibrate severely at high RPM. To be in dynamic balance, the prop has to be balanced evenly for the same reason a car tyre is balanced to prevent it from vibrating at high speed on the road.

There are at least two methods to balance the prop. One, the prop can be removed from the aircraft and placed on a workshop stand for balancing. This is a relatively simple method but not as accurate as using an electronic balancer, such as a Chadwick-Helmuth Vibrex Dynamic balancer. A small accelerometer is placed on top of the engine, which measures vibrations in inches per second at different RPM. The electronic balancer indicates where, and how much weight is to be placed on the prop to balance it. Again, this is similar to balancing a car wheel. A correctly balanced propeller will have the vibrations reduced to around 0.04 inches per second. A smooth running prop makes the ride more enjoyable and less fatiguing for the plane's occupants.

7 – Turboprops, Propulsors & Propfans

Turboprops

Turboprops are of two different types, the free-turbine and the fixed-shaft. The fixed-shaft is also known as single-shaft or direct-drive. This type of engine has its compressor/turbine mounted on the same shaft as the propeller. The other type of turbine is the free-turbine where the compressor/turbine is mounted on its own shaft separate from the prop shaft. It supplies a flow of exhaust gas pressure to the power turbine attached to the prop shaft. There is no direct mechanical link between the power turbine and the gas generator, hence the name of free-turbine. A very good example of this type of engine is the well-known Pratt & Whitney PT6A, which in its various versions powers three quarters of the light turbine fleet in the Western world.

The Beech 1900D commuter airliner has two PT6A-67 free-shaft turboprop engines rated at 1279 SHP each.

7 – Turboprops, Propulsors & Propfans

The turboprop engine, whether it is fixed-shaft or free-turbine, works on the same principle as the jet engine, where most of its heat energy is converted to shaft horsepower by the turbine. As opposed to the turbojet or turbofan which ejects a small volume of air rearwards at high velocity to generate thrust, the turboprop imparts a low velocity to a large mass of air via the propeller. The turboprop therefore, has higher propulsive efficiency than the turbojet at relatively lower air speeds, especially on take-off. The exhaust on some turboprops also ejects a small amount of jet thrust and when this is added to the shaft horsepower (SHP) available from the propeller, it is then termed equivalent shaft horsepower (ESHP). The prop thrust produced with the engine idling whilst taxiing can be more than sufficient and to warrant the use of frequent brake application or beta mode. To alleviate this problem, the constant speed unit may have a ground fine/flat pitch stop to reduce the blade's angle of attack

The Scottish Aviation Jetstream 1 is powered by twin Turbomeca Astazou fixed-shaft turboprop engines. This aircraft is located in the RAF Museum Cosford, England.

to a value less than that of the normal fine/flat pitch stop setting, thereby reducing thrust.

The two different types of turboprop engines require a slightly different approach in handling on the part of the pilot. The free-shaft turboprop is operated in a similar way to the piston-engine where the thrust lever (throttle) controls the compressor speed of the gas generator, while the prop RPM and hence the power absorption, is controlled by the propeller pitch lever. Propeller RPM (N1) and compressor speed (N2) are therefore both controlled independently of each other. Because the compressor turns at a very high speed of around 15,000 RPM or more, a reduction gear is installed to drive the prop at a more sedate speed of 1000–2000 RPM. With the propeller and power turbine separated from the compressor, propeller inertia does not retard the very high acceleration possible with the free-turbine. Although the free-shaft engine has three engine controls, (the third is the condition lever or fuel control) compared to only two controls for the fixed-shaft engine, the free-shaft is a much easier engine for the pilot to operate.

Power output on the fixed-shaft engine is controlled by either adjusting the fuel flow or by varying the compressor RPM and power absorption via the propeller pitch and thus the airflow through the engine. Fuel flow control and the constant-speed unit's selection of blade angle to maintain RPM are interlocked to avoid compressor surge or over temping the engine. A low RPM and high fuel flow will cause an over rich fuel/air ratio, virtually flooding the engine. The engine operates at a very high RPM throughout all flight regimes and due to the narrow RPM range between flight-idle and maximum RPM, a high increase in power will require a rapid change in the blade angle. An engine failure on a fixed-shaft turboprop is far more serious than it is for a free-turbine. This is especially so during the approach when the prop blade is at a low angle of attack (fine/flat pitch) because the prop will absorb the large power required by the compressor. An auto-feathering device is required to turn the blades to the feathered position at as high a rate as possible.

Propulsors

Propulsors have been around for many years now, their origin dating back to 1910. The propulsor is a propeller of short radius mounted inside a duct or shroud, hence its alternate name of Ducted Fan or Shrouded Prop. It is ideally suite to special purpose-built light aircraft or airships, which are designed to fly missions at around 80 knots or slower.

Due to resonance effects, Propulsors always have an odd number of blades, usually around five-blades per prop. The prop blade radius is relatively small around 0.72 times the size of a conventional propeller, but the propulsive efficiency is as good as or better than a normal propeller of greater diameter. The advantage of the smaller diameter of the Propulsor's blades allows it to operate to higher RPM without propeller reduction gear, before the tip speed suffers from compressibility problems. It will produce a similar amount of static prop thrust per horsepower as a prop 1.3 times larger due to the presence of the shroud. The shroud's air intake section is bell shaped, or to be precise, it is an annular airfoil. The shroud is designed specifically to help smooth the airflow through the prop disc. The air's inflow velocity into the shroud increases, resulting in a drop in air pressure in front of the prop. The air flow then exits the duct with the air flow stream tube the same diameter as the duct exit, as opposed to the vena

An Unmanned Aerial Vehicle (Drone) powered by a Ducted Fan. This UAV resides in the Udvar-Hazy Centre, Chantilly, Va.

contracta of a conventional prop. Prop tip vortices will be greatly reduced due to the close tolerance between the prop tips and the shroud. Because the propeller radius is shorter than a conventional propeller, it follows the blade tip speed will be lower resulting in less prop noise and freedom from compressibility problems. In the rare event of the prop throwing a blade, it could be contained within the shroud, a distinct advantage especially on an airship!

Although the Propulsor has some advantages over the conventional prop, it does have some disadvantages too. The shroud will absorb some of the noise making it quieter to observers at the side of the aircraft (it is ideal for airships). However, this advantage is offset by the fact the noise will be focused fore and aft, making it noisier in those areas, which is where the cabin is located on single-engine Propulsor aircraft. The shorter radius of the prop could imply a saving in weight, but this may be offset by

7 – Turboprops, Propulsors & Propfans

the weight of the extra blades (usually five or more). In addition, the weight of the shroud must also be taken into account, plus its extra cost and drag penalty. For all the Propulsor's advantages and disadvantages, only a small increase in efficiency is realized, and that is mostly at the low end of the speed range.

Some helicopters have a Ducted Fan tail rotor known as a Fenestron, a type of Propulsor. The French Aerospatiale's Gazelle (1967) and Dauphin (1972) helicopters both have a thirteen blade, composite Fenestron mounted in a duct within the tail boom for improved performance, replacing the conventional tail rotor. The later 1989 model Dauphin's fenestron has only eleven blades to further improve its performance.

To conclude, the propulsor powered aircraft has its propulsive efficiency suited to the lower end of the speed range. At the high end of the speed range, the propulsive efficiency of the Propfan is suited to the medium range

The Eurocopter EC 130 with a tail rotor
Fenestron, which reduces tail rotor noise by
50% due to its unevenly spaced blades.

airliner, designed to cruise at speeds closer to those of the jet-powered aircraft with turboprop efficiency.

Propfans

The Propfan is a turbine engine driving a tractor or pusher, single or contra-rotating, multi-bladed propeller of advance design. It was intended to be the power plant of the future for short to medium haul air transport aircraft. Turboprop aircraft are ideal for short haul feeder routes cruising at around 350 knots at altitudes up to 25,000 feet, while the jet transports cruise higher at 450–550 knots. The Propfan is designed to combine turboprop economy with jet transport performance. The Propfan's major disadvantage is the fan blade's high tip speed associated with the relatively high cruise speed of the aircraft. The high tip speed suffers from separation of the air flow boundary layer over the blades causing noise and a loss in efficiency due to compressibility problems. More of this later.

From the foregoing, it is obvious the Propfan is not an ordinary propeller, but an advanced turboprop designed to meet a specific need in the air transport industry. The Propfan is used on aircraft, which are designed to cruise at a much higher speed than conventional prop driven aircraft, and must be designed accordingly. The main difference between Propfan and propeller aircraft is the number and shape of the individual blades. The relatively straight shape of normal prop blades is totally unsuitable for the high operating transonic speeds of the Propfan. Therefore, the Propfan's blades are swept back scimitar shape, with the sweepback being far more prominent than a turboprop's scimitar shaped blades. This shape is essential to maintain efficiency at high transonic speeds

A dramatic double-exposure of the McDonnell Douglas MD-80 test-bed with the P&W Allison Model 578-DX Propfan, at sunset.

Photo courtesy Hamilton Standard, Connecticut, USA.

for the same reason a jet aircraft's wings are swept back. The blades are designed as high-speed airfoils with a thin lenticular section and high aspect ratio planform, producing a high lift/drag ratio. A very high helix angle is also used, which combined with its high advance per rev produces in effect, a relatively slow air flow over the blades. Ironically, although transonic speeds require a prop with scimitar shaped blades, a straight blade is required for a fully supersonic propeller, due to aeroelasticity problems of stress and flutter where a straight prop blade is superior to a scimitar blade.

Propfan blades are made from titanium or composite materials to provide greater strength, lightweight and stiffness. It is much easier to attach a great number of lightweight composite blades to a prop hub than heavier metal blades. Six-blade composite props are becoming more common on new generation turboprop transport aircraft, with up to twelve blades being used on a contra-prop allowing smaller diameter props. The Propfan blades appear to be short and stubby due to the hub/radius ratio of 0.45, double that of a conventional propeller. This provides the advantage of lower tip speeds of around 750–800 feet per minute, although some Propfans may have tip speeds operating in the transonic range of Mach 0.8 to 1.2, if the designer requires this. Although it maybe acceptable for a Propfan's blade tip to operate in this range, the rest of the blade is usually operating at the more conventional subsonic speeds. On a conventional prop, smaller blades have a reduced diameter with increased prop disc loading, leading to a loss in efficiency. Adding a greater number of blades, as found on a contra-Propfan for example, reduces the prop blade loading thereby regaining the lost efficiency. The multi-bladed Propfan handles the undesirable problem of compressibility better than a conventional prop. The air flow through the Propfan cascades over the numerous blades allowing compressibility to build up gradually with a minimum loss of energy. Depending on the Propfan's design, the prop disc loading (as opposed to prop blade loading) may be similar or up to twice that of a conventional propeller.

The number of blades, small diameter, scimitar shape and high aspect ratio, all leads to an increase in propulsive efficiency 20% better than a conventional high by-pass turbofan at Mach 0.8. At cruising speeds between Mach 0.7 and 0.85, the propulsive efficiency is about 72–80% for a single-rotation Propfan and 85–90% for a contra-Propfan. This puts the Propfan's efficiency as being better than both a conventional prop and turbofan at the Propfan's design cruise speed of around 450 knots.

Fuel burn will be down 25–45% less than a conventional jet engine, which was the original reason for developing the Propfan due to the fuel crises of the mid 1970s. The fuel burn figures and propulsive efficiency for a contra-Propfan are better than those for a single rotation Propfan. The wake turbulence between the two fans of a contra-Propfan increases noise by 2–3 dB, although the aft fan will recoup some of the lost power in the wake from the front fan. Installing a duct or shroud around the Propfan will attenuate the noise, which being higher than a conventional prop is a distinct disadvantage for a civil transport aircraft. With a duct or shroud installed, the Propfan then becomes known as a Ducted Fan. In the rare event of the prop throwing a blade, it could be contained within the shroud, a distinct advantage especially on an airship!

For all its advantages, the Propfan research ended in the USA, however it continued slowly in Russia with production models now flying.

8 – Constant-speed Units

The Constant-speed Unit

The constant-speed unit (CSU) located at the propeller hub is a speed-sensitive device used to select or maintain a constant engine RPM within the CSU's operating range of approximately 1600–2800 RPM. The unit is one of a variety of types in general use powered either electrically, or more commonly by hydraulics. The advantage of using a constant-speed propeller over a fixed-pitch prop is increased take-off and climb performance due to maximum engine RPM producing greater brake horsepower, a lowering of fuel consumption and less wear on the engine. Power, air speed and air density are three variables affecting the propeller, which are compensated for by the constant-speed unit.

Basically, a constant-speed unit makes use of fixed and variable forces to change the blade angle. The CSU's governor supplies the variable force and the fixed force is either from internal compressed air pressure, or from the aerodynamic forces acting on the prop blades and centrifugal forces acting on the counterweights to oppose the governor. Which way the governor or the forces act depends on the design of the CSU and to some extent on the prop's condition of operation, either under power or windmilling.

The CSU governor is connected to the engine's crankshaft by gearing and so will detect any change in the engine's RPM. Any increase in RPM above the preset value will cause the governor to turn the blades to a coarser pitch. This will place a greater torque load on the engine causing a decrease in RPM to the preset value. The opposite happens with a reduction in RPM; the governor produces a lower blade pitch angle reducing torque and

8 – Constant-speed Units

An AkroTech Giles G202 unlimited-level, aerobatic aircraft.

allowing the engine revs to pick up again. The hydraulic CSU makes use of the engine's lubricating oil acting on a piston to adjust the blade angle. Normal engine oil pressure maybe used, but on some units the oil pressure is boosted by a pump attached to the governor. Increased oil pressure has greater power to adjust the blade angle more quickly. Feathering props usually incorporate an auxiliary oil system to feather the blades because normal engine oil pressure maybe insufficient following an engine failure. The linear movement of the piston is transmitted by linkages, cams and/or gears to rotate the prop blades to the required angle.

The constant-speed unit found on single-engine light aircraft and fixed-shaft turboprops use oil pressure to turn the blades to coarse pitch. These are the oil counterweight type made by Beech and McCauley and are less complex, cheaper and lighter in weight than the type used on light piston-twin aircraft. For twin-engine aircraft and free-shaft turboprops, the CSU works in the opposite sense, with oil pressure turning the blades to fine/flat pitch, favouring the air/oil and Hamilton Standard Hydromatic CSU's. The advantage of this type is when an engine fails; a loss of oil pressure will start the blades moving through coarse pitch into the feathered position.

Purpose-built aerobatic aircraft employ a CSU similar to those mounted on twin-engine aircraft, where the oil pressure turns the blades towards fine/flat pitch. The continuously varying attitudes achieved by aerobatic aircraft can momentarily decrease the lubrication oil from reaching the CSU. If the CSU worked on the principle where the oil pressure increases the blade angle, this could result in a prop over-speed depriving the engine of its vital oil supply; this is also a very good reason to avoid extreme attitudes in non-aerobatic aircraft. Using the CSU

The Yak aerobatic aircraft fleet use centrifugal weights attached to the propeller to assist prop pitch change. This is a YAK 55m model.

type where the oil pressure turns the prop blades to fine/flat pitch ensures the prop reaches its fine/flat pitch-stop before the engine can over-speed or be starved of lube oil.

The following is a brief look at some of the CSU's principle of operation, pilot handling and the faults that may occur in the unit.

The Oil/Counterweight Type

One of the most common type of constant-speed unit is the oil/counterweight type, which consists of an engine driven centrifugal governor with an oil valve and spring loaded counterweight. Oil pressure is supplied to the hydraulically operated piston by a governor driven oil pump which boosts the engine lubricating oil pressure from about 60 PSI to 275 PSI. The desired RPM is selected with the propeller pitch control, which in turn, controls the flow of oil to or from the piston via the governor operated oil valve. Oil pressure then moves the piston in the stationary cylinder to turn the blades to coarse pitch, assisted to a lesser degree by the aerodynamic turning moment of the propeller blades. The opposing force produced by the counterweights will turn the blades towards fine/flat pitch assisted by the centrifugal turning moment of the blades. The aerodynamic and centrifugal turning moments were covered earlier in 'Prop Stress'.

The Beech & McCauley Types

The Beech and McCauley types of CSU operate in a similar manner to the units described above. However, with these types of units there are no heavy counterweights attached to the blades. The prop blades are turned towards fine/

This Cessna 206 Stationair floatplane is powered by a three-blade McCauley propeller.

flat pitch by the centrifugal turning moment and towards coarse pitch by oil pressure acting on the moveable piston.

The Hamilton Standard Type

The Hamilton Standard CSU operates on the oil/counterweight principle. The major difference is the direction of movement of the blades due to oil pressure acting on the piston in the cylinder. The increased pressure moves the cylinder forward over the stationary piston. The cylinder is connected to the blades by a system of cams and gears. An increase of oil pressure turns the blades to fine/flat pitch assisted by the centrifugal turning moment acting on the blades. The opposing force is supplied by the centrifugal force acting on the heavy weights attached to the blade hub and the aerodynamic turning moment acting on, and turning the blades towards coarse pitch.

Note the oil pressure acting on the piston turns the blades in the opposite direction to that described for the Beech and McCauley CSU's.

Air/Oil Type

The air/oil CSU works in the same manner as the Hamilton Standard CSU with the addition of compressed air pressure to assist turning the blades towards coarse pitch. The compressed air chamber is filled with dry air or nitrogen gas at a pressure of approximately 175 PSI. The reason for using dry air is to prevent corrosion and freezing of the moisture within the system. Compressed air is used on newer types of CSU in place of counterweights in order to save weight.

The Hydromatic CSU

The constant-speed units mentioned above are all of the single-action type. The oil pressure acts on only one side of the piston with the opposing force provided by counterweights, compressed air and aerodynamic or centrifugal turning moments. In some units, a spring may also be used to assist the centrifugal turning moment.

For transport aircraft, Hamilton Standard introduced the Hydromatic double-action CSU, where oil pressure is directed to either side of the CSU as required. Normal engine oil pressure is fed to the upper side, or front of the piston to turn the blades to fine/flat pitch, assisted by the centrifugal turning moment. Engine oil pressure boosted to 450 PSI by the governor boost pump is directed to the lower side or back of the piston to turn the blades to coarse pitch. An increase of oil flowing to either side of the piston changes the prop blade angle via the cam connected

The Douglas DC-3 uses Hydromatic double-acting and feathering propellers.

to the piston. The advantage of having a double-action CSU is the need and weight of the spring and counterweight are excluded.

The Curtiss Electric Propeller

The Curtiss Electric Propeller was very popular during WW II, and was used on many American single-engine fighters and light twin-engine bombers.

The Curtiss prop's CSU worked on a different principle to the CSU's mentioned above, using an electric motor to turn the blades either way. The reversible electric motor is driven by power from the aircraft's battery and AC electrical system. The operation of the governor opens and closes circuits to drive the constant-speed unit in the required direction. Through a system of gears connecting the electric motor to the blades, the blade angle is changed.

8 – CONSTANT-SPEED UNITS

The Curtiss P-40 was equipped with an Allison V-1710 engine powering a Curtiss Electric, three-blade prop.

A brake holds the blade angle constant and then releases its hold whenever the electric motor is activated. The fine and coarse pitch limit stops are also electrically operated, with the addition of a mechanical fine/flat pitch limit stop to prevent reverse pitch being inadvertently selected.

It is interesting to note, the Lockheed YC-130 Hercules transport prototype and the first ten production aircraft were fitted with Curtiss-Wright, three-blade electric propellers. Powered by the Allison T56 turboprop engine, the props rotated at a constant 1108 RPM; the thrust was varied by adjusting the prop's blade angle. However, a problem with the electrically operated governor caused the CSU to overspeed or 'hunt' in either direction. This resulted in uneven thrust being produced and caused the aircraft's nose to yaw from side to side. Overheating of the gearbox required frequent engine shut-downs. The problem with the Curtiss-Wright props was eventually corrected but not before a change was made to hydraulic

CSU's made by Aero Products, Allison's subsidiary company. Subsequent Hercules models from the C-130B onwards were equipped with Hamilton Standard three, and later four-blade propellers. However, to follow the trend of modern turboprop aircraft, the new C-130J models have a new type of free-shaft turboprop engine – a Rolls Royce AE 2100 of 4637 SHP driving six-blade composite props.

The table overleaf briefly lists the different constant-speed unit types and their method of operation under normal forward thrust conditions.

Aircraft Class	Type	Fine/flat pitch	Coarse Pitch
Light aircraft & Fixed-shaft turboprop	Oil/counter-weight	C/W & CFTM	Oil pressure & ATM
Light aircraft	Beech & McCauley	Increased oil pressure	CFTM
Multi-engine	Hamilton Standard plus CFTM	Increase oil pressure plus ATM	CFF on counter-weight
Multi-engine, Free-shaft turboprop, Aerobatic	Air/oil type	Increased oil pressure	Compressed air & ATM
Transport Aircraft	Hydromatic	Normal oil pressure plus CFTM	Boosted oil pressure
World War II (fighters & bombers)	Curtiss Electric	Electric motor	Electric motor

Abbreviations used in the above list:
C/W = Counterweights
ATM = Aerodynamic turning (or twisting) moment
CFTM = Centrifugal turning moment.

9 – Propeller Operation

Magneto Drop & Leaning the Mixture

During the pre-take-off engine run-up, the magnetos are checked at around 1700–1800 RPM. At this stage the propeller pitch is at the fine/flat pitch limit stop and remains there during the engine run-up and until after the take-off is well under way. Therefore the drop in RPM will still be apparent and will not be masked by the CSU. The same applies to leaning the mixture if this is necessary before take-off, as would be the requirement when operating from a hot and high airfield. Lean the mixture until the RPM rises to its maximum peak. The exhaust gas temperature (EGT) will also be at its peak indicating the correct mixture setting for maximum power. When the RPM is at its peak, note the position of the mixture control and then quickly return it to 'full rich'. The engine should continue to run smoothly without any surge in power. Continued smooth running indicates the chosen mixture setting was correct and the mixture control can now be returned to the continuously noted position.

Governor Check

After checking the magnetos are functioning correctly, the next item on the check list is the CSU governor. The RPM is increased further into the prop governor range for cycling the prop; 2000 RPM is a normal figure to use. The prop lever is retarded into full coarse pitch where the extra blade drag will cause the RPM to rapidly drop. When this occurs, the prop control must be quickly and smoothly returned to the fine position. The RPM should then return to its original value to indicate the blades are moving

9 – PROPELLER OPERATION

freely throughout the entire pitch range and returning to the fine/flat pitch stop, ready for take-off. If the engine has just been started from cold, cycle the props three times; if the engine is warm, twice should be sufficient. Cold oil in the system, either from standing overnight or from flying at high altitude, will be too thick to operate the system smoothly and the prop may hunt, it will not maintain the required RPM. Hence the need to cycle the prop three times on the first start-up of the day to ensure the thin oil is flowing freely through the CSU. Therefore check oil temperatures are in the green before run-up. During the prop cycling, the prop should respond to the coarse pitch selection within 2–3 seconds of coarse pitch application. If it does not, it requires the attention of a power plant technician.

The next checklist item if the prop is of the feathering type, especially during the first run-up of the day, is to check the feathering system. After checking the magnetos and cycling the prop, the engine is throttled back to 1700 RPM or thereabouts depending on the type. The prop pitch lever is then retarded to the feathering gate where initially there should be no change in the RPM; an RPM change indicates a faulty CSU. Retarding the prop pitch lever through the gate into the feathering position, results in an RPM drop due to propeller drag; the prop blades are at a 90 degree angle to the relative airflow when the aircraft is stationary. The RPM will drop to about 600–800 RPM and the sound of the engine changes from a steady hum to a throbbing sound as the blades turn into the feathered position. With this change in engine sound, the prop pitch lever is then returned to its fine/flat pitch position. The RPM should not be allowed to drop below 1000 RPM because this will place undue stress on the engine. In cold weather, the oil in the CSU may not initially run freely preventing a smooth

PRM drop when performing the feathering check. Repeat the procedure until the RPM drops evenly.

The term feathering is taken from boat rowing where the oar blade is turned parallel to the water's surface as it is returned ready for the next rowing stroke.

Running Square

During normal cruise flight, a power setting of 24 inches Hg manifold pressure and 2400 RPM would be a commonly used figure for modern light aircraft with a normally aspirated (non-turbocharged) engine. This power setting is known as 'running the engine square' or, '24 square' for short. [Knock off the last two digits of the RPM number; in this example it is 2400 minus the last two zeros to equal 24]. If the manifold pressure is increased to a higher figure than the 2400 RPM the engine would be running 'over square'.

Conversely, at a manifold pressure below 24" Hg the engine would be running 'under square'. During the normal operation of a non-turbocharged engine, it is usual practice to run the engine either square or under square. Except for turbocharged engines, running the engine over square is not normally recommended for low time pilots new to constant-speed props.

Running some engines too far over square results in over-boosting which can cause serious damage to the engine in the form of pre-ignition, detonation and high cylinder head temperatures. To avoid this problem a set sequence is recommended to increase or decrease power settings. When increasing power, lead with the RPM/pitch control lever followed by the MP/throttle control. The procedure is reversed for decreasing power; reduce MP/throttle first followed by the RPM/pitch change. However, there is an exception to the rule, which will be discussed shortly.

Reducing Power

When a non-turbocharged engine is running, the manifold pressure will always be less than the ambient air pressure, which is approximately 30" Hg at sea-level. This is due to the piston in the cylinder acting like an air pump. Any change in engine RPM will cause a change in the manifold pressure due to the amount of cylinder volume swept per minute by the piston. This becomes significant when changing power settings. For example, assume a take-off power setting of 2700 RPM and a manifold pressure of 26" Hg is being used and when airborne a power setting of 24" Hg is required. If the manifold pressure is reduced to 24" Hg followed by an RPM reduction to 2400 RPM, the manifold pressure will rise back up about 1" Hg to 25" Hg. This will require a further throttle adjustment to return the manifold pressure back to the required 24" Hg. The initial rise in manifold pressure is due to the decrease in the cylinder volume swept per minute causing less suction through the manifold. To avoid having to adjust the throttle/manifold pressure twice, simply throttle back to a manifold pressure of 1" Hg lower than that required. For example here, throttle back to 23" Hg followed by RPM reduction. The manifold pressure will then rise back up to the required 24" Hg.

A check of the aircraft's flight manual (POH) will show a table for RPM/MP combinations. From the table it can be seen a manifold pressure of 2–3" Hg above over-square is permissible. This can be used to advantage during the initial letdown from the cruise altitude. Reducing power by 200–300 RPM first, instead of reducing MP, will not only reduce noise but it will also increase the MP as already mentioned earlier. Increased MP will produce an increase in cylinder head temperature, counteracting the drop in temperature normally associated with reduced power settings used during the descent. When the RPM has been reduced to its limit in accordance with the POH table the manifold pressure must be reduced to avoid exceeding the engine's operating limits. Also, check the cylinder head temperatures is not exceeded. Finally, after reaching circuit height, reduce the manifold pressure from the descent/cruise setting before selecting full fine/flat pitch for landing; otherwise, the high RPM associated with full fine/flat pitch will cause excessive noise and can be objectionable to people on the ground.

The prop thrust is proportional to the manifold pressure, and RPM/blade angle. A change of MP has the greatest effect on thrust variation with RPM being constant. Conversely, changing the RPM/blade angle has less effect on thrust developed. During an approach to land, RPM is maintained at a fairly high level by using fine/flat pitch, power is varied by throttle use. In the event of a 'go-around', maximum thrust is quickly supplied by increasing the manifold pressure while the CSU turns the blades to a coarser pitch setting to absorb the increased power.

Lock-on

The CSU is a very reliable device but faults can, and do occur. A blocked oil feed line to the CSU will cause the blades to lock-on to the pitch setting in use at the time the blockage occurs. It maybe possible to remedy the fault by cycling the prop-pitch control to clear the blockage. However, if the blockage persists, the flight should be terminated as soon as convenient. Keeping in mind a fine/flat pitch setting can over-speed the prop and a coarse pitch can be a problem if a go-around is necessary after a baulked landing.

Overspeed Condition

A constant-speed unit malfunction in flight may cause the prop blades to move into full fine/flat pitch and overspeed, or run-away. The engine RPM may rapidly increase and exceed the maximum limits. This maybe accompanied by a high-pitch whining sound caused by the very high prop tip-speed. The engine should be throttled back and shut down immediately. If this is not done, the engine may burn up due to failure of the lubricating system in an over-heated engine and followed by a possible engine fire. This can of course lead to the total loss of the aircraft. Alternately, the high centrifugal force on the prop blades caused by the high engine RPM may result in throwing a blade. With the prop now out of balance the engine could be torn from its mounts, with fatal results. However, it may be possible in some cases to throttle back the engine to a low power setting and air speed and land at the nearest suitable airfield.

The WW II, Boeing B17F Flying Fortress bombers were notorious for propeller run-away problems. These bombers were retrofitted with Hamilton Standard paddle-blade propellers and Hydromatic CSU's. The low temperature at high altitude caused the oil to congeal preventing it from flowing freely to the CSU. This resulted in difficulty in feathering the prop, which induced prop run-away problems, causing some engine to be torn from the mounts with the loss of the whole aircraft in some situations. If the prop over speeds on a twin-engine aircraft, the increase of thrust produced by the extra high RPM may cause a yaw towards the good engine, the opposite way to an engine failure. On the other hand, the high RPM may cause a tremendous drop in prop efficiency and thrust. This could cause high drag resulting in a yaw in the usual direction for an engine failure. Therefore, check both engine tachometers and the vertical speed indicator. If an engine fails during the climb, the rate of climb will greatly reduce, whereas a prop overspeed will maintain, or nearly so, the rate of climb near its normal limit. The American FAA certification requirements state in the event of a governor failure the static RPM should not exceed 103% of the engine's rated RPM. This requirement determines the position of the prop's fine/flat pitch stop.

Windmilling

In the event of an engine failure, the prop will reduce speed to around 1200 RPM as it windmills. The power required to cause the prop to windmill is provided by the free air stream flowing through the prop disc. In an attempt to maintain RPM, the CSU will decrease the blade angle to the fine/flat pitch stop. However, a windmilling prop produces more drag in fine/flat pitch than it does in coarse pitch. Therefore, on a single-engine aircraft select full coarse pitch before the engine stops running. The required oil pressure to the CSU piston will be lost once the engine has stopped and then it is too late to select coarse pitch. Reduced prop drag improves the aircraft's glide ratio enabling it to cover a greater distance in the ensuing forced landing. [See Windmilling Prop Forces].

The Feathering Prop

On multi-engine aircraft, the prop can be feathered in order to stop the engine to prevent windmilling drag

9 – Propeller Operation

and any further damage to the engine. Feathering must be achieved quickly before the engine stops, otherwise it maybe impossible to get the blades to feather. Opening the throttle a small amount maybe sufficient to increase the RPM above idle; alternatively, lowering the nose may help to keep prop windmilling long enough to feather it before reverting to safe single-engine speed (V_{MC}). This can only be done if altitude and time permits. An engine failure, on or shortly after take-off requires immediate action to get the prop feathered.

The air/oil type CSU is more suitable for twin or multi-engine aircraft; in the event of an engine failure, loss of oil pressure would allow the opposing air pressure acting on the CSU piston to turn the blades into the feathered position. The aerodynamic turning moment, which normally turns the blades towards coarse pitch, would no longer be of assistance. In fact, it is more of a hindrance due to the force being reversed when the prop is windmilling and it attempts to turn the blades towards fine/flat pitch. [See Prop Stress].

A feathered prop has its blades turned to a pitch angle of approximately 90 degrees edge-on to the air flow to reduce aerodynamic drag and vibration caused by the disturbed slipstream flowing over the wing and tailplane. Due to the blade's twist, only the middle portion of the blade is parallel to the airflow, while the blade's inner and outer portions are presented to the airflow at a positive angle of attack in opposing directions; this will tend to rotate the prop in opposing directions with the net result, the prop remains stationary.

Feathering the prop in flight can be achieved by various methods depending on the design installation. These methods are:

THE FEATHERING PROP

- Manually moving the prop pitch lever through a detent on the throttle quadrant
- The use of a feathering button to activate an electro-mechanical pump, or
- An auto-feathering system.

Some of the smaller and lightweight composite props employ a blade counterweight system to assist the pitch change mechanism. When an engine fails, the counterweights will automatically cause the blades to turn towards the feathered position instead of their natural tendency to turn towards fine/flat pitch.

The pilot should be familiar with the feathering and un-feathering procedure for his/her aircraft. If an engine is shut after a real failure, it is the usual practice to leave it so and carry out a single-engine landing as soon as possible. The decision to restart an engine after a failure should not be taken lightly, because of the risk of fire, further engine damage, or the inability to re-feather the prop again if a re-start is not possible.

It should be kept in mind when feathering a prop for training purposes, in cold weather the oil in the CSU may become congealed quickly. Congealed oil can impair the operation of the CSU and present difficulties in un-feathering the prop again at the end of the exercise.

Un-feathering the prop is a relatively simple procedure that can vary between different aircraft types. It also depends if an oil pressure accumulator is used for un-feathering the propeller. For a given type and model of aircraft, some have accumulators and others do not. It is imperative to know the correct procedure for the aircraft. The accumulator holds oil under pressure and when activated oil is directed to the CSU to un-feather the prop.

9 – Propeller Operation

This is a one-time method only, so if it does not work correctly the first time you could be stuck with a feathered prop until after landing.

After selecting the coarse pitch for less prop drag or fine/flat pitch if an oil pressure accumulator is used, the engine starter is engaged. Oil pressure returning to the CSU as the engine comes back to life, will move the blades out of the feathered position. With the slipstream acting on the prop blades, the engine will start easier than it does on the ground. However, expect a fair amount of vibration until the engine has returned to active duty at the normal engine RPM. Un-feathering can also be achieved on some aircraft by activating the feathering button to start the auxiliary pump, which will supply oil to the CSU. Un-feathering requires greater oil pressure (around 600 PSI) than that required for the initial feathering. As the prop blades move out of the feathered position, the air flow through the prop disc due to the plane's air speed will start the prop windmilling. When the RPM passes through a pre-determined figure of around 800–1000 RPM, the feathering button is activated as the CSU returns to its automatic operation. Holding the feathering button in for too long will cause the blades to move through to the fine/flat pitch stop and cause damage. Hence, the need to deactivate the feathering button. Because there are different methods of feathering and un-feathering the prop, depending on the aircraft type, a full knowledge of the particular system and procedure is essential.

An engine shut down in flight will cool rapidly. It will need some time to warm again at a low power setting after re-start before opening up to cruise power; check oil temperatures and pressures and the cylinder head temperature gauge. The throttle should be set at the recommended manifold pressure and the prop control moved from the fine or coarse pitch setting, whichever was used for the un-feathering procedure.

Ground Feathering

When shutting down an engine in flight, the centrifugal force produced by the windmilling prop holds open a spring loaded 'start lock' allowing the blades to move through the coarse pitch stop into the feathered position. However, when shutting down the engine on the ground, the spring will override the decreasing centrifugal force and close the start locks as the prop speed decreases through 800 RPM on piston-engines, thus preventing the blades from feathering on shut down.

On aircraft with free-shaft turboprop engines, such as the P & W PT6A, the propellers always park in the feathered position as opposed to the propellers of a fixed-shaft

A CASA CN 235 transport parked with the
props in the ground feathered position.

9 – Propeller Operation

turbine engine and on most piston-engines, which park in fine/flat pitch. The free-shaft turboprop does not have or need start locks; the blades will turn to the feathered position when the engine is shut down. During start-up, the starter has only to spool up the turbine/compressor. As the gas pressure builds up, the gas generator and prop will spool up in there own time. When the RPM builds up sufficiently, the prop will automatically move out of the feathered position into fine/flat pitch.

The advantages of feathering on shut down are prolonged windmilling is avoided and the wind will not blow the prop around when the aircraft is parked, due to lack of compression to prevent it. This will reduce danger to personnel around the aircraft. A prop brake maybe installed (known as an arrested prop system) to stop the prop's rotation when the engine is left running during a quick turn-around between trips.

On fixed-shaft turboprops, it is impossible to stop the prop rotating because the compressor/turbine is both mounted on the same shaft. Because the compressor/turbine and propeller all turn as one unit, the propeller is parked in fine/flat pitch to reduce prop drag on engine start-up. A pitch setting other than fine will cause excessive prop drag that would retard the engine acceleration to normal idle speed resulting in a hung start. To hold the prop blades in fine/flat pitch, centrifugal start locks are incorporated in the design. Part of the engine start sequence on some turboprops is to release the start locks by selecting reverse thrust with the power levers and then returning them to the fine/flat pitch setting.

Negative Torque System

Under certain flight conditions with the engine throttled back to idle and the prop in fine/flat pitch, the prop may produce zero thrust, or drag. The drag in this condition is known as the 'flat plate drag'. Imagine the prop disc as being a large solid plate. Turboprops with their large diameter props are particularly susceptible to drag under this condition, the fixed-shaft turboprop being affected more than the free-shaft turbine engine. The CSU would naturally select a fine/flat pitch setting, but this produces the highest drag and negative torque with the engine idling. To alleviate the high drag and negative torque at low power, turboprop engines employ a negative torque system incorporated within the CSU/reduction gear assembly. The negative torque system senses when negative torque is being produced by the prop and commands the CSU to turn the blades to a more coarse pitch setting. At this setting the blade angle will be correct to absorb a predetermined amount of horsepower. As soon as the negative torque is removed with increased power selection, the CSU resumes normal operation.

Autofeathering

Autofeathering is a feature more common on large transport aircraft, particularly turboprop transports. A turboprop with a reduction gear of around 20 to 1 will have its engines driven up to a speed of twenty times the prop speed. This could place enormous stress on the transmission between the engine and propeller and cause excessive prop drag. With the engine being driven by the prop, the negative torque sensor will activate the autofeathering system and feather the blades automatically.

9 – Propeller Operation

The autofeathering system operates on a different basis to the negative torque system, which senses when the engine is still running under power but producing negative thrust, as opposed to an autofeathering system that senses a failed engine condition and turns the prop blades into the feathered position. The negative torque system operates on a continuous basis 'as required' at low power settings. Autofeathering acts only once – when the engine has failed.

Simulated Zero Thrust

It was mentioned earlier, pilots of multi-engine aircraft should be proficient at shutting down the engine and feathering the propeller in flight. Some training maneuvers require simulating an engine failure and this should be done at a safe altitude where possible. At low altitudes such as just after take-off, any training advantage gained in shutting down the engine are far out-weight by the risks involved if mishandled. Therefore, to avoid having to shut down the engine completely, the throttle can be set to simulate zero thrust. If the engine is throttle right back to the idle position, the prop will produce greater drag than when it is feathered, opening the throttle slightly to produce about 11 inches HG manifold pressure, prop drag will be overcome. The actual power setting will vary between each aircraft type and this can be found in the aircraft's flight manual (POH). With the throttle slightly open, the prop will produce a small amount of thrust equal to the prop drag. A state of balance will then exist with the net result, zero thrust will be simulated.

Reverse Thrust

Following an engine failure, the CSU will decrease the blade angle to fine/flat pitch in an attempt to maintain RPM as the propeller's rotational velocity decreases. If the prop were the reversible pitch type, the blades would go into reverse pitch if it were not for the fine/flat pitch stop or the autofeathering system, if installed. A squat switch on the undercarriage, or release triggers on the throttle, or some other method is used to remove the fine/flat pitch stop in order to allow the prop blades to move into reverse pitch. Needless to say, reverse pitch should never be selected until the aircraft is firmly on the ground, due to the possibility of an excessively high and dangerous rate of decent. If one prop fails to return to forward thrust, a severe asymmetric condition will result. This has actually happened with fatal results.

Full throttle produces full thrust but zero thrust is produced with the throttle slightly open. When the throttle is fully closed, a small amount of negative thrust or prop drag is present. With the aircraft firmly on the ground, reverse thrust is applied by retarding the throttle through the gate or detent to idle reverse. This action will turn the blades via the CSU to a fixed negative blade angle of around 30 degrees past the fine/flat pitch stop. At idle reverse, the prop will produce about 60% of the maximum reverse thrust available. Further retardation of the throttle lever, will power the prop to a higher RPM than reverse idle to produce the total amount of reverse thrust available. Full aft movement of the throttle lever produces full reveres thrust caused by the increase of engine power absorbed by the prop. Reverse idle may provide sufficient braking force on long runways without the need to go to full reverse thrust. A turboprop produces about one third of its maximum shaft horsepower (SHP) when full reverse is

9 – Propeller Operation

applied, reducing the landing roll by about one quarter to one third of the unassisted reverse thrust landing distance. Another advantage of reverse thrust is the fact it destroys the wing's lift placing more weight on the undercarriage wheels for increased braking. The disadvantage here is the degraded elevator effectiveness; all the wheels should be firmly on the ground to prevent the nose-wheel dropping on quite.

To achieve maximum benefit of reverse thrust it should be applied fully and early as possible in the landing roll. It is more effective at higher speeds just after touchdown than it is when the landing roll is nearly complete. This is due to the fact, the aircraft's kinetic energy is destroyed quicker and the reverse thrust force is greater due to the addition of the aircraft's forward velocity. As the aircraft decelerates, the prop blade's negative angle of attack reduces with the result, the reverse thrust also reduces. Reverse thrust should be cancelled when the aircraft's ground speed is down to around 40–50 knots. However, there are some exceptions to this rule; check the aircraft's flight manual (POH). Reverse thrust should be avoided if possible when operating on unsealed airstrips due to gravel damage to the blade's leading edge and foreign object damage to the engine.

It is possible to taxi the aircraft backwards using reverse thrust, but the brakes should be used with caution to prevent the aircraft tipping on its tail. Visibility behind the aircraft is also a problem.

Safety Around the Prop

The above paragraphs have covered propeller handling inside the cockpit; this section deals with handling the prop and moving around near the prop while on the outside of

Safety Around the Prop

the aircraft. A rotating prop is an almost invisible blur and hard to see and hear when other aircraft are operating nearby. Safety around the prop is paramount at all times no matter if the prop is turning or stationary. The following are a few points to keep in mind when moving around the aircraft before and after flight. This maybe stating the obvious but accidents do happen that can be avoided.

The propeller per-flight involves the following:

- Check for leading edge knicks, cracks or dents
- Check the security of the prop spinner; it should be firmly secured in position
- Any signs of oil or fluid leaks require a more detailed inspection, which can be difficult if a prop spinner is in place. A leaking CSU can drain oil from the engine sump leading to an engine failure.

An accumulation of grass and dead bugs on the blades can lead to an unbalanced prop, plus a loss in prop efficiency due to the rough blade surface. A wipe over with an oily rag would solve this problem, along with an occasional polish with auto wax. Oil or dirt on the blades can be removed with carbon tetrachloride or a solvent; check the flight manual for the prop manufacturer's recommendations for cleaning the prop blades.

The propellers should never be used to man-handle the aircraft on the ground because this can place undue stress on the propeller and CSU. Holding onto a prop to pull or push the aircraft places bending stress on the prop blades. Maneuvering the aircraft by holding the prop at the blade root places stress on the CSU, the crankshaft and bearings. Whenever the aircraft has to be moved, always use the tow bar, recommended handgrips or taxi the aircraft under its own power. However, if you insist on using the prop, it is

9 – Propeller Operation

This Victa CT4B Airtrainer has a very distinctive black and white stripe on the forward side of the prop for better visibility.

better to push or pull at the blade roots or the prop boss and not on the spinner, keeping in mind what was said above.

Before the first flight of the day, it is usual practice to turn the prop by hand to clear the engine cylinders of 'hydraulic lock'. 'Hydraulicing' as it is known, is caused by incompressible engine oil draining into the lower cylinders of a radial engine preventing rotation during engine start-up. With the horizontally opposed type of engine, hydraulic lock is not a problem; therefore, it is not necessary to turn the engine over by hand. Nevertheless, if you do decide it must be turned over, then turn the prop in the direction of normal rotation and never backwards, as damage maybe caused to the engines ancillary equipment, such as the vacuum pump, etc. The term 'pulling through' or 'hand turning' are applied when the prop is turned by hand for reasons other than starting the engine.

A few simple rules apply to swinging the prop by hand to start the engine:

Safety Around the Prop

- Check the aircraft is parked into wind with the brakes 'on', preferably with the wheel chocks in place with the tail pointing away from the hanger and other buildings or aircraft

- The plane should be sitting on a smooth, firm surface, so you won't slip over when you swing the prop

- Do not wear any loose clothing that may get in the moving prop

- Keep both feet firmly on the ground, one foot in front of the other so you can quickly step backwards after each swing of the prop, incase the engine starts.

Some pilots may dispute this last method having their own variation on the theme. However, what ever method is used, hand-propping is now a dying art and caution should

This DH Beaver has white prop tip stripes on its grey prop. The DH Chipmunk to its right also has a high-visibility prop.

9 – Propeller Operation

prevail. If you have any doubts about swinging the prop, seek help from an experienced pilot or flight instructor. Besides, an experienced pilot should be at the controls any way. Hand propping solo, has on many occasions allowed the aircraft to taxi away and collide with other aircraft and buildings, etc, or become airborne with no one at the controls.

A constant-speed prop can be moved fore and aft by a small amount; this movement is allowable and is known as 'blade shake'.

Always treat the prop with care and attention it needs and deserves. In addition, always check the mags are off and the throttle closed before touching the prop. It can be a lethal device if the engine starts unexpectedly, so not touch it unless absolutely necessary. Pilots and passengers alike have a fascination for propellers and having their photos taken while hanging onto a prop blade. Imagine the result if the engine turned over! Your passengers should be warned of the dangers of the propeller and be escorted by you to and from the plane. Do not allow people to board or de-plane with the engine running Stepping off the leading edge of a low-wing plane when the prop is turning is asking for trouble because the prop is difficult to see from that position. The prop is an almost invisible disc when viewed from the rear of the plane and there are not many rotating devices in an unguarded state in any other industry. A rotating prop is as dangerous as a butcher's ham slicer. The prop can slice off an arm or a leg quite easily. Ensure the prop is stopped before anyone gets near it.

Prior to engine start, it is essential to ensure the area around the aircraft is clear of all personnel and that no one is in danger of walking into the rotating propeller. This author always used the pre-start check "all clear, front and rear" to ensure no one is in danger of walking

SAFETY AROUND THE PROP

A prop warning line on a Beech C-54 (Model 18) Expeditor. This aircraft is located in the National Museum of the USAF, Dayton, Ohio.

into the rotating prop. The old method of calling "Clear prop" before starting is still a good point of airmanship, but hardly ever used these days. Turning on the rotating beacon prior to start-up also helps to warn people of your intention to start.

Military aircraft usually have markings – prop warning lines – on the forward fuselage to warn of the location of the propeller. However, manufacturers of civilian aircraft are reluctant to post warnings on their aircraft; is it for aesthetic reasons? They could help prevent someone from walking into the rotating propeller. On single-engine aircraft, the words "Prop" and arrows, pointing forward towards the prop, could be painted on the engine cowl. This could be a good safety feature on aero club/flying school and personal aircraft, where less experienced pilots are operating these aircraft and taking friends along for rides.

Safety around the prop is the pilot-in-command's responsibility at all times, either from inside the cockpit or outside the aircraft.

Conclusion

The workings of an aircraft propeller involve many variables. Some factors work together to enhance the prop's efficiency while other factors oppose efficiency. The result is a propeller suitable for one aircraft may not be suitable for another. Propeller design has seen many changes over the years; changes in blade planform and tip shape and also changes in materials used in their manufacture. All these changes enhance the efficiency of the aircraft propeller. How will the design of future props change? And what benefits will they have over present day props? Today's props generate greater thrust more efficiently than the props mounted on aircraft of years gone by. Their performance has improved tremendously over the years. Therefore, we can expect to see some new and interesting innovations in prop design in the future as new aircraft types are brought onto the flight line.

However, it is not only new aircraft types that are benefitting from the new prop designs, because new props are now being deigned for individual aircraft types both old and new. It is no longer necessary for the aircraft designer to choose 'off the shelf' propellers as was once the case. The prop manufacturers are refining the prop's design, so now props with three blades producing better cruise speed performance on aircraft which were once powered by two-blade props.

Conclusion

There is no doubt the prop is an ideal thrust generator for aircraft flying in the lower speed range, below 350 knots and will always remain so. Since the early 1960s, the jet engine has reigned supreme as the prime mover of large transport aircraft. With the advent of the Propfan design in the later half of the 1970s, will the propeller go full circle (pun intended) to make a comeback and once again power the medium to large aircraft of the future's airlines? Maybe not. With more than half the world's aircraft powered by propellers driven either by piston-engine or turbine power, the propeller powered aircraft is going to be with us for many years to come.

Glossary

Activity factor. A measure of the quantity of power a prop can absorb.

Actuator disc. Froude's theoretical propeller disc associated to the axial momentum theory.

Advance/diameter ratio. Symbol 'J = V/ND'. The ratio of the aircraft's forward speed to product of the prop RPM and diameter.

Advance per rev. The distance the propeller advances forward in one revolution. Equal to the effective pitch.

Aerodynamic turning moment. A turning force acting on the pitch change axis caused by the aerodynamic loads on the prop.

Airscrew. An alternate name for the propeller usually refers to a tractor propeller. The word airscrew is now archaic.

Arrested prop system. A prop brake used on free-shaft turboprop aircraft.

Asymmetric blade effect. Uneven thrust over the prop disc caused by the down-going blade being at a greater angle of attack with the prop axis inclined upwards. Also called asymmetric disc loading.

Auto feathering. The prop feathers automatically after an engine failure.

Automatic reverse pitch. Reverse pitch for braking in the landing roll; can be armed in flight and activated by microswitch on touchdown.

Axial momentum theory. Rankine-Froude's theory on how propellers produce thrust.

Axial thrust. Propulsive thrust produced by the propeller.

Beta angle. A negative blade angle relative to the plane of rotation.

Beta range. The range of blade angles from fine/flat pitch through reverse pitch.

Beta mode. The use of the Beta range of pitch angles.

Blade angle. The angle between the blade's chord line and the plane of rotation. Associated to the geometric pitch.

Blade back. The blade's back corresponds to the upper surface of an aircraft's wing. With a tractor prop, the back of the prop is seen when viewed from in front of the aircraft!

Blade cuffs. Increased area of blade root, which increases prop efficiency by 1–3% and reduces prop noise by around 3 dB. Increases air flow to engine intakes. Also known as an extended blade root.

Blade element. A thin section of blade corresponding to an airfoil section.

Blade face. The relatively flat surface corresponding to the wing's lower surface. Also known as the thrust face.

Blade loading. Brake horsepower divided by prop blade area.

Blade profile. The shape of a cross section of blade.

Blade section. The shape of the blade element.

Braking pitch. Negative blade angle to provide reverse thrust for braking.

Breakaway thrust. Extra thrust to start the aircraft moving from a stationary position.

Cavitation. A near vacuum at the prop tips on the suction face of the blade caused by very high tip speed.

Centrifugal turning moment. Component of centrifugal force causing a turning moment about the pitch change axis towards fine/flat pitch.

Chord line. The chord line is an imaginary straight line joining the leading and trailing edges of an aircraft wing or propeller blade.

Coarse pitch. A high blade angle used at low RPM.

Constant-speed propeller. A prop with a CSU which automatically maintains a chosen RPM within the prop governor's range.

Constant-speed unit. A governor system installed in the hub of a constant-speed propeller.

Contra-rotating props. Two co-axial props mounted and driven, usually by the same engine, but rotating in opposite directions. Contra-props or co-axial prop are alternate names.

Counter-rotating props. Props mounted on two different engines and rotating in opposite directions.

Design air speed. The aircraft's forward speed at which the prop produces maximum efficiency.

Design point. A combination of forward speed and RPM at which the propeller provides maximum efficiency.

Discing. The use of ground fine/flat pitch for braking purposes.

Ducted Fan. A short radius prop with an odd number of blades encased in a shroud. Also known as a shrouded prop.

Efficiency. A measure of how well the propeller converts engine torque into thrust, expressed as a percentage.

Effective pitch. Equal to the advance per rev; related to the helix angle.

Effective pitch ratio. It has the same meaning as advance/diameter ratio. Also called slip function.

Effective prop thrust. An alternate name for net-thrust.

Electric propeller. An electrically controlled constant-speed unit.

Experimental pitch. It is equal to the advance per rev when the prop produces zero net thrust.

Exponential mean pitch. An alternate name for experimental pitch.

Feathered prop. A condition where the blades are turned to a pitch angle of 90 degrees edge-on to the airflow to reduce drag.

Fenestron. A multi-bladed ducted fan, helicopter tail rotor.

Fine/flat pitch. A small blade angle of 2–3 degrees used for take-of and landing. Also known as flat pitch.

Flat pitch. See fine/flat pitch, above.

Free fan. Boeing's name for their Unducted Fan.

Froude's efficiency. See ideal efficiency.

Geometric mean pitch. The mean pitch of all blade elements from root to tip.

Geometric pitch. The distance a prop advances in one revolution when the blade's angle of attack is zero degrees. Related to blade angle.

Ground adjustable prop. A variable pitch propeller, ground adjustable only.

Ground fine/flat pitch. An ultra-fine/flat pitch to provide braking after landing,

Handed props. Left, or right-handed props refers to the direction of rotation of tractor props when viewed from the rear of the aircraft.

Helix angle. The angle between the relative air flow and the prop's plane of rotation.

Hotel. A reference to using an arrested prop system.

Hot prop. A prop with electro-thermal anti/de-ice system.

Ideal efficiency. Prop thrust related to the aircraft's speed and propeller inflow factor. Also known as 'Froude's Efficiency'.

Ideal pitch. Also known as the zero thrust pitch, exponential mean pitch, or experimental pitch.

Induced inflow. The low pressure area of propwash drawn into the front of the prop disc.

Induced outflow. High pressure area of propwash behind the prop.

KTAS. The true air speed in knots.

Left-handed prop. See 'handed prop'.

Negative thrust. Prop thrust acting in the reverse direction of the plane's motion.

Negative torque system. A system that senses when negative torque is generated by the engine and adjusts the blade angle to a more coarse setting. Propwash Velocity V. KTAS

Nominal pitch. The geometric pitch at a nominated propeller radius, usually the 70% radius.

Nose cap. A small type of propeller spinner.

Paddle blade. A wide chord, low aspect ratio blade.

'P' factor. See 'asymmetric blade effect'.

Advance/diameter ratio. The ratio of propeller pitch to diameter.

Pitch distribution. The changing blade angle along the length of the blade.

Plane of rotation. The plane in which the blade tips travel.

Power differential. The difference between thrust horsepower available and thrust horsepower required.

Pressure differential. The difference in pressure between the front and rear of the prop.

Prop blade loading. Engine brake horsepower divided by prop blade area.

Prop blast. An alternate name for propwash.

Prop disc area. The total area swept by the prop blades.

Prop disc loading. Engine BHP divided by the prop disc loading.

Prop face. See 'Blade face'.

Prop fan. An advanced turboprop engine driving a multi-bladed propeller of short radius.

Prop overspeed. Prop RPM exceeds limits after a CSU failure.

Propulsive efficiency. The efficiency of the aircraft's propulsive system.

Propulsive thrust. The net propulsive force driving the aircraft after subtracting the various losses from the free thrust.

Propulsor. An alternate name for ducted fan.

Propwash. An alternate name for prop blast.

Pusher prop. A propeller mounted aft of the engine.

Race rotation. Rotational velocity imparted to the propwash by the prop. Also known as helical velocity.

Reduction gear. A gear system used to reduce the prop's speed below engine RPM.

Reverse pitch. Prop blades at a negative angle of attack to provide reverse thrust.

Right-handed prop. See 'handed props'.

Rotational velocity. The speed of prop rotation. Also known as tangential velocity.

Shrouded prop. An alternate name for ducted fan.

Slip. The difference between the advance per rev (effective pitch) and the geometric pitch. Also expressed as a percentage of the difference between the propwash velocity behind the prop and the aircraft's velocity.

Slip function. The ratio of the aircraft's velocity (TAS) to the product of prop RPM times the diameter (V/nD). See 'advance/diameter ratio'.

Slipstream. The air flow passed the aircraft and unaffected by the propwash.

Solidity. The ratio of total blade area to total prop disc area.

Speed ratio. The ratio of aircraft forward speed to propeller RPM.

Spinner. A streamline fairing over the prop hub.

Standard pitch. The geometric pitch measured at the 75% standard radius, or at another specified location.

Static prop thrust. Prop thrust produced at maximum RPM when the aircraft is stationary.

Stress raiser. Alternate name for prop nicks.

Synchronizer. A device to match RPM of both or all propellers.

Synchrophaser. A device to match the position of blades on all props.

Thickness/chord ratio. The t/c ratio is approximately 4% at the blade tips and increases to about 25% at the blade root.

Thrust-apparent. Free thrust with interference of engine nacelle or fuselage taken into consideration.

Thrust differential. The difference between the thrust available and thrust required at a given power setting.

Thrust-dynamic. Prop thrust equal to the air mass moved per second times propwash velocity minus the aircraft's velocity.

Thrust face. See 'blade face'.

Thrust-free. Propwash thrust when unaffected by the body behind the prop.

Thrust-gross. The propwash-thrust when disturbed by a body behind the prop.

Thrust-net. Gross thrust minus drag of a body behind the prop.

Thrust/torque ratio. The greatest amount of thrust for the least amount of torque.

Tip speed. Tangential velocity of prop tips; forward velocity is ignored.

Tractor prop. A propeller mounted in front of the engine.

Trailing vortex drag. Alternate name for induced drag.

Variable-pitch prop. A prop with two or three adjustable pitch positions.

Velocity of advance. Equal to propwash velocity but ignoring prop tip RPM. Greater than aircraft forward speed while under engine power.

Vena contracta. The position in the propwash where the contraction and velocity are greatest.

Windmilling prop. A condition of the prop when driven by the free air stream flowing through the prop disc.

Zero lift line. Related to a given negative angle of attack and experimental pitch at, which the prop produces zero thrust (or lift).

Zero torque pitch. Advance per rev when windmilling with zero torque.

Bibliography

Collier, b, *The Airship.* Grenada Publications, 1974.
Gibbs-Smith, Charles H. *The Aeroplane – An Historical Survey.* HM Stationary Office.
Gibbs-Smith, Charles H. *The Invention of the Aeroplane.* Faber & Faber.
Combs, C.F. & Morgan, E.B., *Supermarine Aircraft Since 1914.*
Bowers, P.M., *Unconventional Aircraft.* TAB Books, 1984.
Taylor, J.W., & others, *The Lore of Flight.*
Sutton, Sir G., *Mastery of the Air.*
Van Sickle, *Modern Airmanship.* Van Reinhold, 1957.
Milne-Thomson, L.M., *Theoretical Aerodynamics.* Courier Dover Publications, 1957.
Stinton, D., *Design of the Aeroplane.* Grenada Publications, 1983.
Kermode, A.C., *Mechanics of Flight.* Longman Scientific & Technical, 1987.
Sweetman, B., *Aircraft 2000.*
Gunston, B., *Jane's Aerospace Dictionary.* Jane's Publishing, 1980.
Heywood, J.E., *Light Aircraft Inspection.*
Durand, W.F., *Aerodynamic Theory.*
James, D.N., *Gloster Aircraft since 1917.*
McMahon, P.J., *Aircraft Propulsion.* Pitman Publications 1971.
Hoerner, S.F., *Fluid Dynamic Drag.*
Jones, L.S., *US Fighters.*

Index

A

Activity Factor 84–85, 93

Aerodynamic Theories 5–6, 96

Air/Oil CSU 175, **178**, 182, 190

Airscrews 1–2

Aspect Ratio 82, **91–93**, 105, 135, 138, 171–172

Auto Feathering System 35, 166, 191, **195–196**, 197

B

Balanced Prop 161, 199

Beech CSU 174, **176–177**, 182

Blade Angle 19, 37, 42–47, 54, 59–61, 65–66, 148, 151, 165, 173–175, 178–180, 187–189, 195–197

Blade Shape 5, 22–23, 32, 37–39, **45**, 49, 103–108, **137–140**, 159, 169, 171

Blade Twist 45, 56, 67, 190

C

Cause of Noise 26, 94, 127, **131–132**, 136–140, 142–143, 143–145, 167–169, 172, 187

Constant-Speed Prop **8–11**, 22, 54, 56, 59–69

Constant-Speed Unit 8–11, 19–20, 40, 60–69, 108, 138, 164–165, **173–181**, 183–184, 187–199

Contra-Rotating Prop 4, 11, **14–19**, 29, 32–34, 124, 129–130, 169, 171–172

Counter-Rotating Prop **14–19**, 125, 126–129

Curtiss Electric Prop 10, 14, 19, 60, **179–181**, 182

D

De-ice **20–21**, 22, 92, 110–111

E

Effect on Aircraft Stability 14–15, **113–130**, 161

Efficiency 4, 13, 26–28, 32, 37, 51–57, 63–65, **71–111**, 118, 133–140, 153, 164, 166–169, 169–172, 189, 199, 204

Experimental Pitch 42, **45–47**, 49–53

F

Feathered Prop 10, 20, 22–23, 61, 67, 122, 148, 150, 158, 166, 174–175, 179, 184–185, 188, **189–193**, 193–194, 196

First Airborne Props **2–3**, 6

First Turboprops 17–20, 22–23, **24–26**

Fixed-Pitch Props 2-4, 11, 24, 31, 40, 42, 47, **56-58**, 69, 72, 94, 107, 127, 142, 143

Forces

Cruise Flight 93, 103, **147-148**, 155

Reverse Thrust 151-153

Windmilling 127, **148-151**, 153-155

G

Geometric Pitch 42-43, 45, 47-51, 51-54, 57-58

Governor Check 183-185

Ground Feathering 193-194

Gyroscopic Effect 121-124

H

Hamilton Standard CSU 10, 175, **177-178**, 182

Helical Propwash 118-119, 130

Helix Angle 42, **43-44**, 45, 53, 74, 148-150, 171

High-speed Aerodynamics 132-135

Hurricanes 11, 28

Hydromatic CSU 10-11, 175, **178-179**, 182, 188

I

Icing 21, **108–111**

L

Leaning 183

Lockheed C-130 **19–20**, 89–90, 180–181

Lock-on **187–188**

M

Magneto Drop 183

Manufacturing 23–24, 138

Materials 13, 23, 72, **137–140**, 171, 204

McCauley CSU 22–23, 174, **176–177**, 182

Minimum Control Speed **125–126**, 150, 190

P

'P' Factor **119–121**, 123–124, 127, 129

Pitch **37–69**, 116–118, 123, 151, 153–155, 159, 165–166, 173–176, 183–192

Power Absorption 16, 67, 69, **82–84**, 85, 94, 165, 197

Prop Blade Drag 5, 57, 67, **103–108**, 133, 183

Prop Blade Loading 71, 88–89, **93**, 131, 155, 171–172

Prop Diameter 37, 54–56, 58, 71, 81, 82, **85–86**, 88, 94–95, 108, 123, 129, 131, 135–136, 140, 143–144, 166, 171–172, 195

Prop Disc Loading 84, **94–96**, 113, 119, 121, 127, 129, 171–172

Propfans **32–34**, 74, 76, 86, 140, **169–172**, 205

Prop Location 37, 101–102, **114–118**, 124, 127, 131

Prop Manufacturers 7–8, **21–24**, 32, 58, 125, 138

Prop Torque Force **113–114**

Propulsive Efficiency **74–77**, 79

Propulsors **31–32**, 74, 76, 132, **166–169**

R

Record Breakers 19, **28–30**

Reducing Power **186–187**

Reverse Thrust 6, 8, 35, 69, 151–153, 155, 194, **197–198**

Running Square **185–186**

S

Safety 156, **198–204**

Simulated Zero Thrust **196**

Slip 42, 49, **51–52**

Solidity 37, 40, 71, 84, **85,** 86, 88, 93, 95, 105

Spitfires 11, 17, 28, 113–114

Stress 4, 93, **153–160,** 161, 199

Supersonic Props 26–28, 30, 171

Synchronizing 142–143

T

Terminology 39–42, 69, 104

Tip Noise 137–140

Tip Shape 140, 204

Tip Speed 26, 37, 74, 94–95, 131, **135–137**, 143–144, 166–167, 169, 171

Turboprops 8, 13, 18–20, 23, 24–27, 30, 67, 74–76, 110, 115, 127, 140, 142–143, **163–166,** 171, 174, 180, 193–194, 195, 197

V

Variable Pitch 59–69

W

Windmilling 127, 148–151, 153, 155, 173, **189,** 189–190, 192, 193–194

Wright Brothers 3–4, 14, 24, 114

Also available from Frank E. Hitchens

Range & Endurance

Fuel-Efficient Flying in Light Aircraft

Range & Endurance — Fuel Efficient Flying in Light Aircraft was written for pilots flying light-single or twin piston-engine aircraft at the Student, Private or Commercial Pilot levels. Using the fuel carried on the aircraft in an efficient manner will not only save money but also increase the aircraft's range (distance flown) or endurance (time remaining airborne).

This book discusses various factors in the efficient use of the fuel available, describes fuel technology, light aircraft fuel systems, refuelling procedures, pre-flight planning in regards to fuel use and in-flight use of fuel to increase the aircraft's range or endurance. The book ends with a final chapter containing fuel calculation formulas for use on the pilot's E6-B Air Navigation Computer. Flying for range or endurance is an important part of a pilot's airmanship duties; this book offers a good insight to achieve this on every flight.

Also available from Frank E. Hitchens

Formulas for the E6-B Air Navigation Computer

Using the E6-B Simply & Efficiently

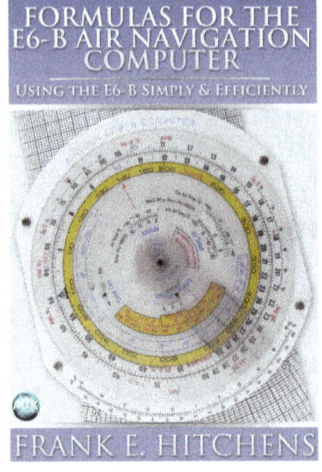

Formulas for the E6-B Air Navigation Computer is written for pilots and air navigators at all levels of experience from the novice to the professional. The book is self-help on how to use the E6-B Air Navigation Computer.

An E6-B Air Navigation Computer is a circular slide rule with a wind slide on the reverse side. It is dedicated to performing all calculations related to pre-flight planning and in-flight air navigation. Every pilot has an E6-B Air Navigation Computer, which is supplied with a very brief instructional booklet when the E6-B is purchased. However, the booklet only covers a few basic formulas, and many more formulas are required for passing the pilot navigation exams at various levels and, of course, for all operational flying. Obtaining all these different formulas from various sources is time consuming, as this author has discovered over the years. They are not readily available in one book.

The formulas are written as they appear when set up on the E6-B Air Navigation Computer. A full description on how to solve each formula is included, along with a worked example and also the methods for using the wind slide to calculate wind triangle and other navigational problems associated with the wind slide. The book is easy to follow by the novice pilot and a convenient reference source for the more experienced pilot. The book is complete with all the formulas a pilot of any level should need to know. It is laid out in a simple way with over 122 formulas and methods, covering Time, Speed & Distance, Air Speed, Altitude Navigation, VNAV, One-in-Sixty Rule, Wind triangle Calculations, Wind Finding methods, Fuel Calculations, Pressure Pattern Navigation and more.

www.ingramcontent.com/pod-product-compliance
Lightning Source LLC
Chambersburg PA
CBHW060950230426
43665CB00015B/2139